EXECUTION MADE EASY

PRACTICAL TIPS FOR LARGE SCALE PROJECTS

BY YOGESH KUMAR
BE (HONS.) AERO, MS MECH

abenav designs

WWW.ABENAV.COM

EXECUTION MADE EASY

Copyright © 2021 by YOGESH KUMAR

Imprint: Independently Published

Printed and Distributed in India by Brand Inspire (OPC) Pvt Ltd.

All rights reserved. No part of this book may be used or reproduced in any manner whatsoever without written permission except in the case of brief quotations embodied in critical articles or reviews.

Book and Cover design by Abenav Designs
ISBN: 9798592932786
First Edition: Jan. 2021

"

From time immemorial, India with its rich heritage, achieved success in many areas of much diversified nature. We had a very glorious past. Time has come to rebuild that glory.

This book captures some important management practices which would be of immense help towards execution of major programmes which our country is likely to launch on infrastructure development for achieving a five trillion-dollar economy by 2025.

The practical tips provided for fast execution are based on personal experiences of the Author who spent about forty years in steering several prestigious aircraft development programmes at Hindustan Aeronautics Ltd. (HAL); most notable being the Light Combat Aircraft (LCA) – **'Tejas'**.

The book will be a useful guide for our aspiring Designers, Managers, and Executives.

To my magnificent parents Sh.Jaikishan Dass and Smt. Angoor Lata, my loving children Manik, Nidhi, Surabhi, Sujit, Saurabh, Radhika, and my grandsons Tarush and Kabir. To my wife Kavita who inspired me to write this book during the COVID period and gave me full support. To my brothers and sisters, their children, all my relatives, friends, and well-wishers who have always been a great source of encouragement.
This book is dedicated to you all.

I also acknowledge with thanks the valuable feedbacks received from my children regarding the Book. I am particularly thankful to my daughter Surabhi for suggesting the title and some changes in the formatting of contents.
God Bless you all.

PREFACE

This book is a sequel to my previous book titled;

"Lead and Execute – The Art of Managing Large Scale Projects", launched in Dec-2014. That book primarily dealt with a vital subject of Project Management. (Ref. 1 in the end).

For a long time, I had been thinking as to what the most important tool of project management was; why was it that in spite of the best efforts put-in, we ended up with time overruns and associated cost overruns! Why did we fail in bringing the product to the customer in the shortest possible time! After lot of thinking, I got an answer that, it was nothing but **'Execution'**, meaning thereby, the way we executed our projects. Therefore, the idea struck me that I should write separately, only on **'Execution'** based on my personal experiences, and hence this book. The Covid period also gave me some free time to crystallize the ideas into a coherent text which I am presenting before you through the medium of this book.

I personally feel that time could not have been more appropriate than today when our great country is targeting a five trillion US Dollar (USD) economy by 2025. This would require build-up of massive infrastructure almost to the extent of 1.25 trillion USD. If we have to achieve the goal as stated, an order of magnitude more focus will be on how we **execute** our major programmes.

From time immemorial, India with its rich heritage, achieved success in many areas of much diversified nature from materials to mining, astronomy to aerospace, medicines, pure sciences, textiles and building technology etc. We had a glorious past. Maybe we were very focused with aligned vision and emphasis on **'execution.'** Time has come to rebuild the past glory. I sincerely hope that the practical tips presented in this book will give some direction towards that.

Dear Readers, Management is still more an art than a technology. It is very much individualistic; it depends on the manager's style, his beliefs and disbeliefs. Hence whatever has been described in the book should be read with that perspective in mind.

I do not claim to be a management expert. Based on my experiences in successfully handling three major programmes in Hindustan Aeronautics limited (HAL) Bangalore, I have brought out a few practical tips, which to my understanding can be helpful for aspiring youngsters, designers and practicing managers in **execution** of their projects.

Although the book is written with Indian audience in mind, the practical approach to **'Execution'** will apply everywhere.

Since the theme of the book is **'Execution'**, I have given the title;

"**Execution Made Easy** – *Practical Tips for Large Scale Projects*".

It will give me immense pleasure if this book can provide some benefits to our Managers, Designers, and their Organizations.

'Jai Hind'

Yogesh Kumar

Bangalore, India

Sep-21

CONTENTS

Chapters

1. India – A Country to Reckon With
2. Execution-The Theme
3. Execution Made Easy – Practical Tips
4. And the Results!
5. Conclusion

Acknowledgements
About the Author
References

Reminiscences

Photographs of VVIP visits to HAL
Photographs of Previous Book Launch

CHAPTER ONE

India – A Country to Reckon with

"India is an idea whose time has come".

India today, has reached a prominent place in global scenario. With 1.35 Billion people, no country, irrespective of its strength, can ignore the voice of India anymore. Our plus points are, over five thousand years of history and an extremely strong cultural and scientific/academic heritage. We gave so many firsts to the world which are known facts. People of Indian origin are occupying top positions in many multi-national companies and even in governmental administration all over the world. In fact, Innovation and Ideas have always been in Indian genes.

With all this, every one of us has to be proud to be Indian.

Take-off stage

India today is well placed to be the leader and economic power in the expected next technological revolution.

Following are the key strong points which can assume prominent roles in propelling our nation;

- Value system
- Well-structured education system
- Strong cultural and scientific heritage
- Ability to work hard

However, as every dark cloud has a silver lining, I have always said, every silver lining also is surrounded by patch of dark clouds.

Though we may be in the take-off stage; an order of magnitude more effort will be required if we have to achieve the status of global player and economic power in all sectors of science, technology, industry, medical, pharma, administration, humanities, and social services etc. to name a few. Imagine, we are number 48 in the list of countries who are doing innovation as per a study of 2019. Therefore, most important would be;

(a) Widening of Research and Development (R&D) base i.e.,

- Fundamental research in next generation technologies; venturing into the fields hitherto unexplored/unknown.
- Applied research with an eye on its application in products and services.

This would mean much higher percentage of GDP spending on R&D; of the order of two percent. Currently, we spend just 0.7

percent as compared to developed countries such as USA, Israel, South Korea, Japan, who all spend 3 to 4 percent. In absolute terms, we are far too low. No wonder; since we are unable to provide enabling platforms, we end up with problem of brain drain. Engineers, Doctors, Scientists of Indian origin are doing wonders abroad. They are a big pool of talent and if we can provide stimulating platforms, they themselves will become anchors of our growth. I have full confidence in our young generation; they have excellent ideas; all they need is leadership and mentoring.

(b) Synergies between various pockets of excellence for maximum benefits; currently we tend to work in **silos.**

(c) Institute–R&D Lab–Industry tie-up with focus on bringing technology and product from Lab to the Field in the shortest possible time. This is most important if we have to grow.

To achieve all that stated above will require massive build-up of infrastructure including human resource and launch of a large number of R&D based, innovative and technology driven programmes. All these initiatives will then lead to an innovation based eco-system which will propel our growth.

While these will form the core of our development plans, the most important will be the operating part, the **'Execution'**, which will play a very significant role.

This book primarily deals with **'Execution'**; as a central theme as I personally feel that is where in spite of our grandiose plans, we fall far short of achieving our goals; both in terms of time frame and cost.

Before I go on further, I would like to quote what our former Hon'ble President Dr. S. Radhakrishnan had predicted almost sixty years ago;

"We will again be able to establish ourselves as some of the pioneers of the human race, not only in metaphysics and spiritual wisdom but even in the scientific, industrial and technological aspects as well".

-Extracts from his speech delivered at Jodhpur University in August 1962.

"India is the cradle of human race, the birthplace of human speech, the mother of history and grandmother of legend, and the great grandmother of tradition".

-Mark Twain

CHAPTER TWO

Execution-The Theme

> *"Execution is the ability to mesh strategy with reality, align people with goals, and achieve the promised results."*
>
> *-Lawrence Bossidy*

Dear Readers, this book is dedicated to the core theme i.e., **'Execution'**. Have you ever realised how important is the part played by **'Execution'** in any project! It is by far the most significant.

You would have noticed in your daily life that road is built, left-out material is still lying by the side for weeks together; the building is complete, the building material is left on the roadside, and so on. Then there are issues related to time overrun and cost overrun. They can all be related to **'Execution'** one way or the other.

What is 'Execution'

I would define **'Execution'** as a **"Process to implement the project from start to finish"**. Word finish is very important here; what it means is that the project should be considered as complete only when it is successfully operationalized and put to regular use. That is my understanding of total **'Execution'**. However generally, as I mentioned earlier, we take a short cut and end-up with one or combination of the following;

- Poor **'Execution'**
- Faulty **'Execution'**
- Part **'Execution'**
- In-effective **'Execution'**

We all think rather simply, that **'execution'** of a task requires only a task-team leader and a set of people. Believe me, it is not so. Therefore, what is it which is over and above this; and that is what this book brings out in simple words in the Chapter that follows. Although the title of the book says, **"Practical tips for large scale projects"**, these are generic and are applicable to all projects; small, medium, or large.

Whatever I have written is drawn from my personal experiences of successfully handling three major programmes in aerospace. Each of these programmes was unique in terms of management drivers, technology drivers, organizational drivers, design objectives etc. as briefly described below;

(a) Light Combat Aircraft (LCA) – 'Tejas'

This was a technological marvel with more than forty disciplines and over hundred work-centers; a challenge in Technology and Project Management. On top of it, we had complete technology denial from USA at the most crucial juncture of the development programme.

(b) Intermediate Jet Trainer (IJT)

The objective here was to go digital in design work-flow process and execute the programme in shortest possible time without compromising safety.

(c) Jaguar Avionics' Upgrade

This was the largest programme ever undertaken by HAL in concurrent design and line production. Here the challenge was to follow a totally indigenous route for development which was earlier planned to be done under design collaboration with a reputed original equipment manufacturer (OEM) abroad, without affecting the delivery schedules of line production aircraft.

Dear Readers, I am sure you will appreciate the diversified nature of all these programmes throwing major challenges in **'Execution'**.

CHAPTER THREE

Execution Made Easy – Practical Tips

"The heights by great men reached and kept were not attained by sudden flight; but they, while their companions slept, were toiling upward in the night."

-HW Longfellow

'**Execution**' is the central theme of this book as the title suggests.

In my opinion, it is the key to success in any project/task; as you will appreciate, an idea is only as good as its execution. We may have grandiose plans, but they all fade out in memory if not executed properly; both in time-frame and cost. **'Execution'** in this context is meaning to cover full life cycle; i.e., from **concept-to-delivery-to-operationalization**. It is a complete chain; one weak link can mar the whole project.

Here, I would like to place before you **Charts shown later vide Figs. 3.1a, b, and c** which are indicative of practical and innovative ideas that got evolved while executing major projects mentioned earlier and were found to be very useful by me.

Believe me **Dear Readers**, if these are followed in letter and spirit, you can be reasonably sure of success in your venture. So, follow the charts as much as possible, and feel for yourself how **"Execution is Made Easy"**.

However, before I proceed to explain the Charts, I want you to keep in mind that **'Execution'** on its own cannot be successful unless the **Executor**, the **Team-Leader**, draws his/her strength and inspiration from;

- Project-mode Management
- Human Resource

Therefore, we shall briefly discuss these before we proceed further.

Project–mode Management

On a macro level, Project-mode Management is the key to successful **'Execution'**. What it essentially means is that the task with all its activities is to be completed in a project-mode scenario. Treat **'Project'** to be your Boss, no one else; all that you do should be for the sake of the **'Project'**. This should apply inclusively to everybody involved in the **'Project'**.

Actually, **'Project-mode Management'** is an art. It requires thought, self-discipline, and will power. It is encompassing and covers a very wide spectrum as shown in diagram below;

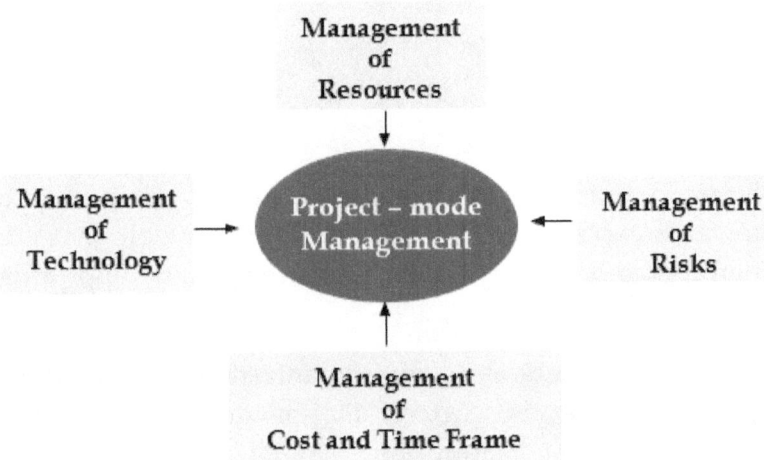

Please remember, Project is like a moving train, it should not be derailed for any reason. Therefore, the team–leader, while executing a project, has to keep an eye on various aspects as shown. Each one of them has to be addressed concurrently.

> "Project Management is an important component of economic development and prosperity".
>
> -Dr. APJ Abdul Kalam*
>
> Former President of India

I was very closely associated with Dr. Kalam for about four years while executing India's most prestigious, the Light Combat Aircraft (LCA) 'Tejas' Programme. Dr.Kalam was then Scientific Adviser to Defence Minister and responsible for development of LCA.

Human Resource

I would put **'human resource'** as a very significant asset towards execution of a project/task. Please note that a large percentage of effective management is dealing appropriately with people.

As is the situation in any organization, the break-up of competency levels through the overall strata of human-resource would generally be;

- Top to high level performers - 20 percent
- Middle level performers - 60 to 70 percent
- Bottom level performers -10 to 20 percent

While the task-team leader will endeavour to train, motivate, counsel and try to push people towards higher categories, those in twenty percent class (top to high performers) have to be nurtured and rewarded as these are the ones largely responsible for success of the project/task.

On this subject, I would like to make a point. I am convinced that everyone who has come for work, has some capability in him/her. This requires competence mapping of each of the members in the team to see which type of job each is best suited to. The task-team leader has to find that out and accordingly make use of it in allocating tasks.

Believe me, if you do this, you will get astounding results. I am saying this from my personal experience of handling directly or indirectly a work force of about one thousand people. Many of my Group Leaders used to come and tell me of this problem; and my advice always was on similar lines; and after a year or so, they would come back and tell me that it really worked. The ability to select the right people and then to delegate effectively to them is the only route to the success of a project.

Also, it should be remembered that harmony amongst the team is very important. It is only then that we can pool together the talents and skills of a lot of people.

Therefore, ignition of spark in everyone is the key to forming a focused and cohesive team.

Remember what Swami Vivekananda said,

"There is a spark in every human being, you should know how to ignite it".

However, while building a team, following aspects have to be kept in mind;

(a) Discipline

This is something which we generally lack. Many of our problems stem from indiscipline and casual attitude. We should get rid of them; otherwise in today's competitive environment, we will be out of race.

(b) Integrity

Everyone in the team should have high level of integrity. We should build this culture in the organization. There will be pressures (internal and external), but the team and its leader should discard all of them and move on. Remember;

"Goal is not important; the route you take to reach the goal is important. Transparency at all levels is the key."

(c) Competency v/s skill

There is a subtle difference between a skilled technician and a competent technician. A skilled individual attains competency only when he/she exhibits a positive mental attitude; one of the most important attributes.

(d) Trust

This is a very important part of effective human resource management. The leader/manager has to have full trust in his/her team members and vice-versa. Then only he/she will be able to get un-solicited support from each one of them.

> *"Good human relations not only bring great personal rewards but are essential to the success of any enterprise."*
>
> *-J R D Tata*

Having dwelt on above two important attributes, (Project – mode Management and Human Resource) of successful execution, I shall now place before you practical tips based on my personal experiences.

For better understanding, I have segmentized the whole process of **'Execution'** in three distinct categories as described below.

A. Execution made easy - Macro level (Fig. 3.1a)

Under this, I have included;

- Leadership
- Relationships
- Technology

B. Execution made easy - Operational part (Fig. 3.1b)

This being the operational part, it is most important and therefore, it comprises of a number of elements as shown in the Chart at **Fig.3.1b**.

C. Execution made easy - Review, Monitoring, and Control (Fig. 3.1c)

Under this are included the following;

- Review Mechanism
- Project Manager

However, please note that all the elements shown above are intertwined and hence have to be addressed concurrently during the process of execution.

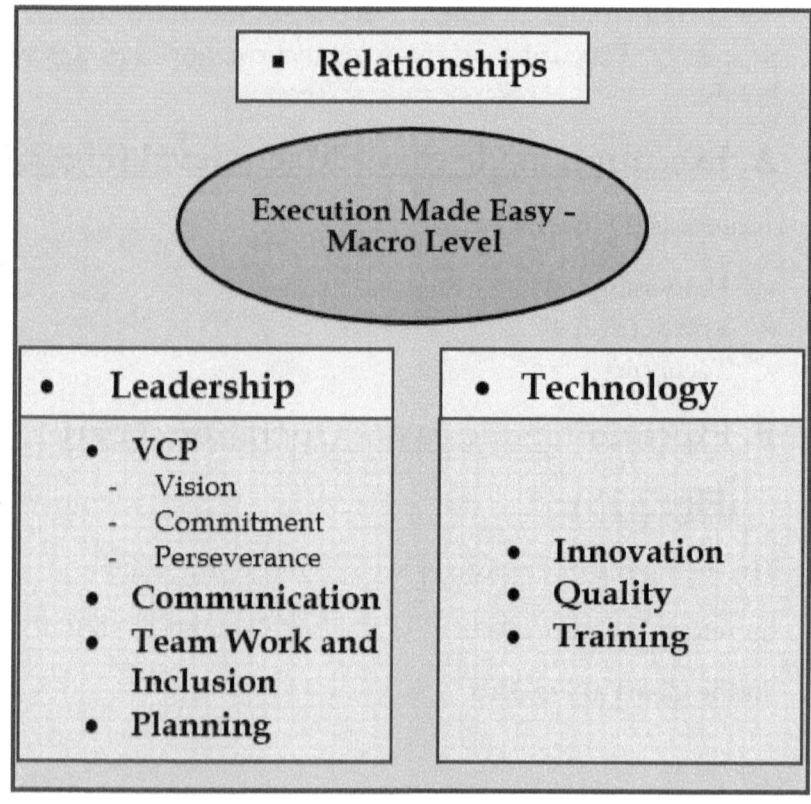

Fig. 3.1 (a)

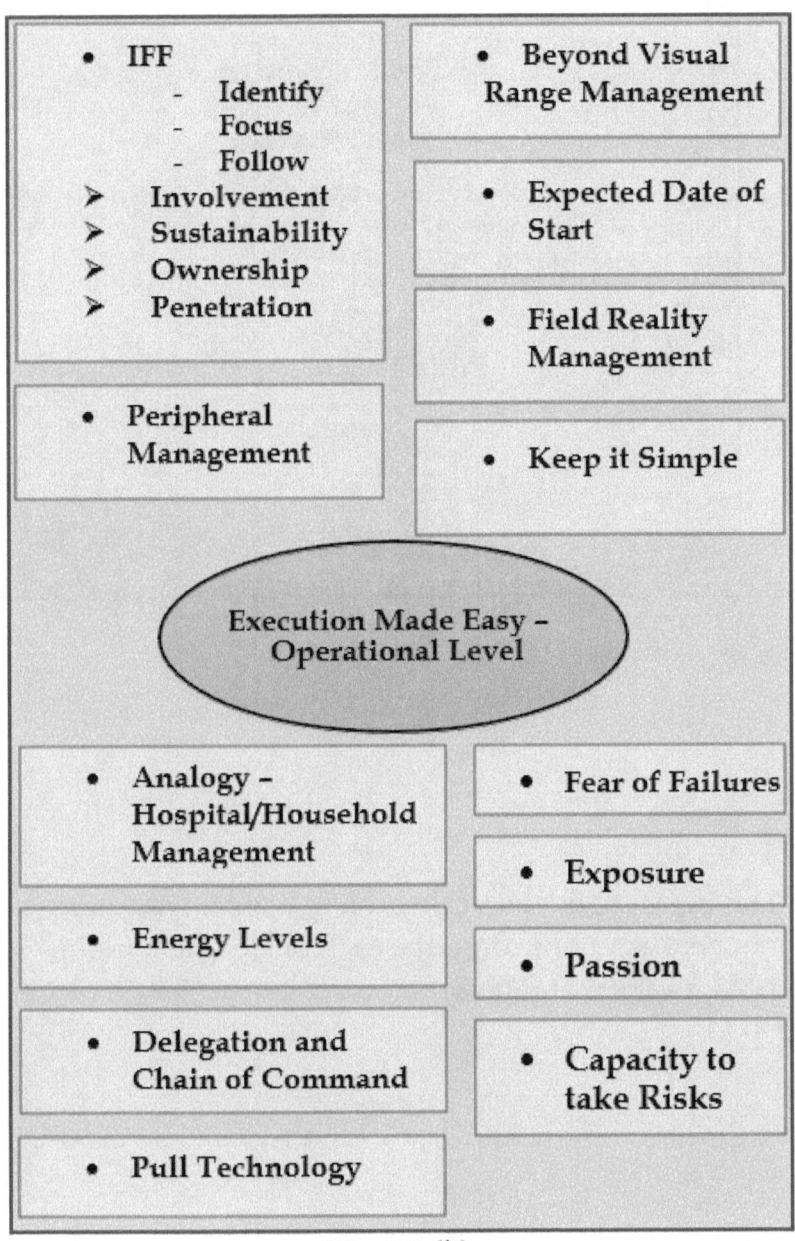

Fig. 3.1 (b)

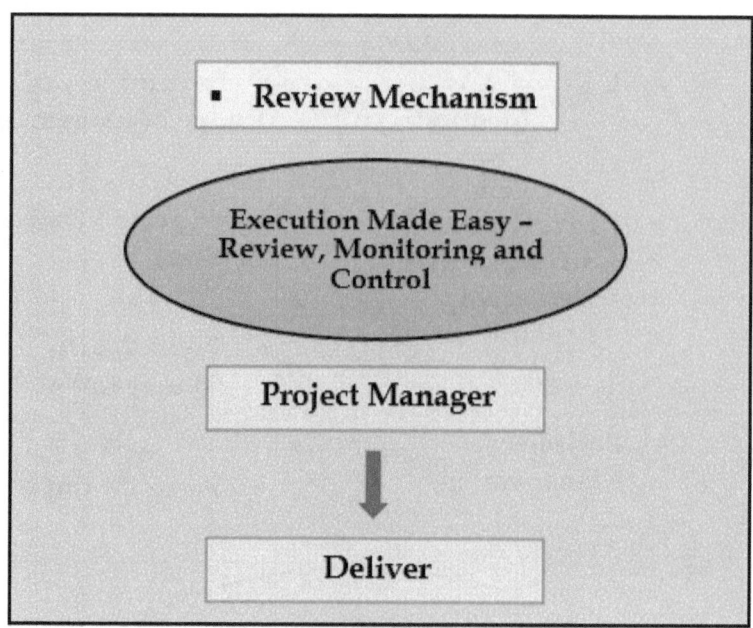

Fig. 3.1(c)

Having categorized various elements of **'Execution'** as above, I shall now describe each one of them in the paragraphs that follow.

A. Execution Made Easy – Macro Level – Fig. 3.1(a): Leadership, Relationships, Technology

(a) Leadership

Leadership plays a very important role in executing a project. It is therefore, a **verb and not a noun.**

First of all, a Leader has to be a role model to whom everybody else looks to for direction, guidance, and inspiration. Lot depends on what he does rather than what he is. Refer to the diagram below;

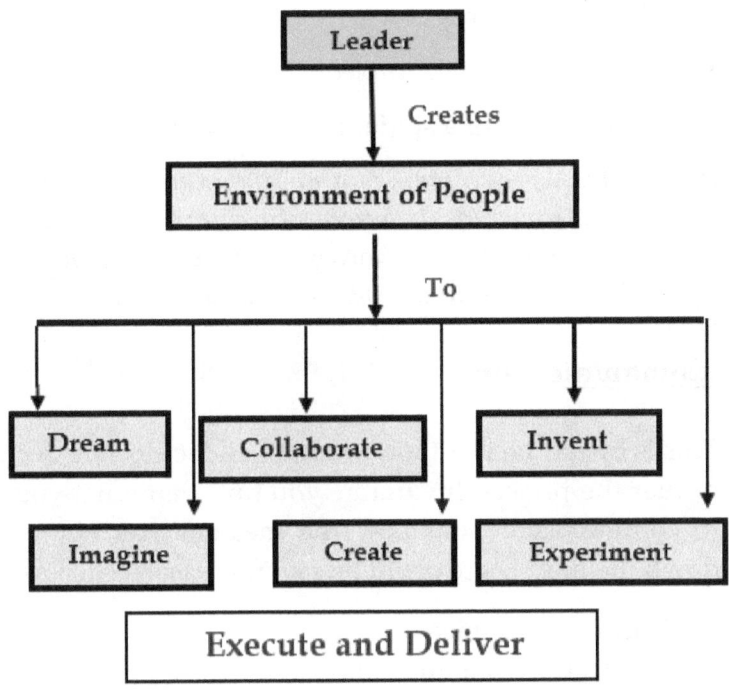

I don't intend to describe the qualities of a good leader as there are number of books available on the subject.

I would rather like readers to pick up some qualities of leadership from a number of live examples that I have given later to illustrate the point. That would be more meaningful.

However, I want to highlight a few important points in the paragraphs that follow.

Vision, Commitment, and Perseverance

No **'execution'** can ever succeed unless the Leader and his/her team have;

- Vision of the project
- Commitment to the project
- Perseverance of a spider to implement the project

In fact, LCA programme is a unique example of all the three above on the part of top aeronautical fraternity, the customer, and the sanctioning authority of our country in those days (80s). The nation will be ever grateful to them.

Communication

This is by far the most important of all factors which can make or mar the project. Invariably you find that what you wanted to communicate, somehow the message has not percolated down the line. There could be several reasons such as;

- Faulty communication
- In-effective communication
- Poor communication

- Inconsistent communication and so on

Take an analogy from our Armed Forces. There, the chain of communication right from commander to the soldier never breaks down. Somehow, the message without any distortion reaches the lowest level; otherwise, the whole system will collapse. **Can we not learn from them!**

Proper Communication is very vital; not only in bonding the team together, but also in aligning the team members towards a common goal. Therefore, improving the communication skills should be a continuous on-going activity and this can only be achieved by training or mentoring or both.

Teamwork and Inclusion

The leader/manager is only as good as the team he/she leads/manages. Therefore, the success of a project depends on teamwork. We have to bring everyone in the mainstream. Many a time it is noticed that while generally people tend to align with the project and its aims and objectives, some percentage is either ignorant or indifferent. These people also have to be brought on-board and that is **'Inclusion'**. Please remember, **indifference is worse than inaction.**

Here, I want to bring-in the concept of **'Stratification of Human Resource'**. You will appreciate that everyone can't do everything; some are good in theoretical work, some are good on shop floor, some have a penchant for field activities like marketing and sales etc. The team leader has to recognize this and allot the task accordingly. This will bring-in **'Inclusion'** and therefore, coherence in the team. For this to happen, he/she has to build a culture where everybody's contribution, direct or indirect is acknowledged and respected. The day this happens, there will be a quantum leap in the team's output.

Remember, **there is spark in everyone; only the intensity differs.**

Planning

Planning is the mother of all programmes. It is known in management parlance that **only that task can be executed well which was planned well.**

This requires a very focused and forward-looking group which has the knack of going into details. So always plan ahead; it is time worth spending. **Also, have contingency plans; if Plan-A does not work, Plan-B should be in place and so on. Build enough capacity to take on surprises.**

Leaders of Tomorrow

Leaders of tomorrow will be required to spend more time in connecting, communicating, and listening, and not in commanding. Their role will be like that of **'Missionaries'**.

While closing on the subject, I would like to suggest that the leader should ask himself this question every day;

"What can I, this day do, to contribute towards excellence in my work!"

During my career spanning about forty years where I had the opportunity to work with larger-than-life personalities, I built in my mind a concept of good leader as the one;

"whose presence is always noted and absence always felt".

Just think deeply about it; you will realize it is very true.

In this context, let me make one more point;

Although for a Leader what he does is important, equally important is what legacy he leaves behind. Leaders will come and go; organizations they belonged to will remain. Therefore, every Leader should create more leaders.

"The beauty of music lies not only in sounds but also in intervening silences. Great musicians leave songs behind; Leaders should be like that".

(b) Relationships

Dear Readers,

I can tell you that **"Relationships"** play an extremely important role. They are even more important when we deal with multi-organizational and multi-disciplinary projects. This is something which many of our leaders and managers tend to ignore and then end up realizing this when it is too late. No wonder that today, many CEOs are being advised to spend a significant portion of their time on relationships and human resource development.

Various Aspects of Relationships - (See diagram below)

Considering the leader/manager as the central point, I have tried to put across following forms of **'Relationships'** as shown in diagram;

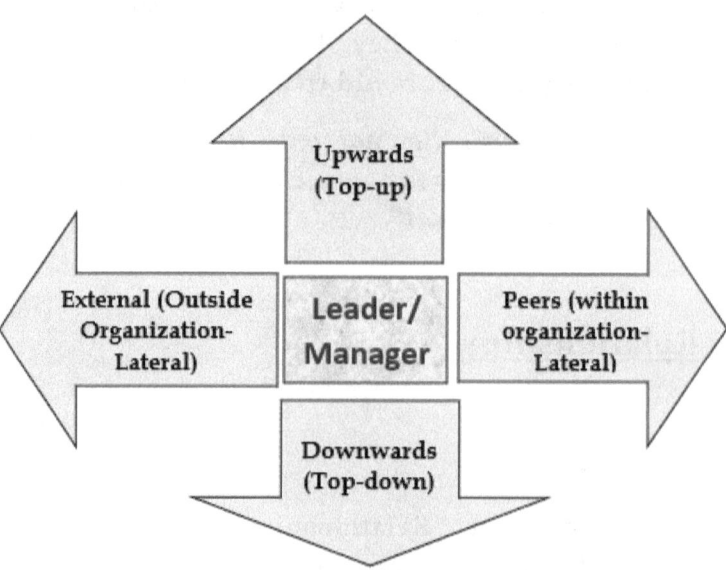

360 Degree relationship

As would be evident from the diagram,

- Upwards (Top-up) means relations the Leader/Manager develops with his/her seniors including the boss.

- Downwards (Top-down) means relations with his/her juniors right up to the workmen and technicians.

- Peers (within organization lateral) means the relations a Leader/Manager develops with his/her counterparts within the organization.

- External (outside organization lateral) means the relations a Leader/Manager develops with professionals related to his/her areas of activities in other organizations.

This is what I call the **"360 Degree Relationships"**. Each of the links shown is important and has to be addressed concurrently.

Few Takeaways

From my own experience, I give below a few **'Takeaways'** which will help every leader/manager in building up robust relationships.

- Go for holistic approach in building relationships. You may have to walk a little beyond convention to understand the people you deal with; whether they are your seniors, or your peers and counterparts, or your workmen, technicians, and employees. Their psyche and temperaments need to be understood. This is what I call holistic approach toward building relationships.

    This takeaway is particularly relevant in respect of people working for you. These are the people who make the organization what it is through its products and/or services. For them, you need to go an extra mile to enhance your engagement, to understand their problems and find solutions; even help them in their personal matters so that when they come for work, they come with free mind to give their best. You have to follow the armed forces where the commander and his men are together in the trenches to fight the war. You have to build a culture where you as a leader/manager will take pride in the work of your team-members before taking pride in your own work. The day this happens, you will notice quantum leap in your team's output. Although difficult, yet it is achievable.

During my tenure, I as a leader, made an unwritten rule that personal matters would get topmost priority provided their redressal would lead to better productivity; and I would say it worked very well.

- In the human psyche, appreciation plays a major role. Everybody needs appreciation for the effort put-in. A leader/manager therefore, should not be restrained in uttering a few words of praise whenever situation arises. It will motivate your work force. One very important point; while you execute your task with a firm hand, be soft at heart, (example of a coconut).

- Invariably it is seen that there is lack of harmony between the team leader and his/her immediate junior (number two). Each one starts by-passing the other; the level of mistrust keeps rising and the whole system collapses. My humble suggestion; have a good working relationship. You may have differences; but it is the team leader who has to take the final call. That has to be respected and honored. This is how our Armed Forces work.

Please remember what I said in the beginning, that your actual **Boss** is the **Project** for which you are toiling day in and day out. But since we all work within the boundaries of an organization, there has to be a well-defined and prescribed disciplinary protocol which we need to follow. Therefore, forget the differences, if any, and work towards execution of the project. This is my sincere advice to all.

- We have to remember one thing when we are working on major programmes. There would be criticisms and negative reporting. Your job, as a leader, should be to take it all in your stride and insulate your team from such environmental factors. Your motto as a leader should be; **"Credit is yours; debit is mine"**. Also remember, as a leader, **you are riding a tiger, you can't afford to fall off the tiger.**

A leader should never criticize the system he inherits or his predecessor once he takes over. This is very common in our organizations where the first thing a new incumbent does is to start criticizing his/her predecessor. Please understand that when you take over, you inherit the entire system with its assets as well as liabilities. It is up to you how you encash the assets and overcome the liabilities.

Remember, time is kind to great people, performers and achievers.

Now I shall narrate a live example of what **'Relationships'** can achieve for you.

This I am taking you back to the time frame of 80s when I was the Chief Designer (Mechanical Systems) at Hindustan Aeronautics Ltd. (HAL) Lucknow.

One of our production test-facilities which used to test the fuel control unit of a reheat system in the fuel factory had a high-speed gear box of very high gear-ratio. It was an imported unit. The rig was the single facility operating 24/7.

During one test-cycle, for some reason (probably lack of lubrication), suddenly the gearbox failed, and the facility went under breakdown. It was a huge setback for our Division as it was affecting the delivery of the aircraft from Aircraft Division at Bangalore who was our customer. We were in the middle of the production year and there was no solution in sight. The situation was serious.

An emergency meeting was called by the Managing Director (MD) the same day to discuss the recovery plan to bring the facility back into operation at the earliest. All Department Heads and General Manger (GM) of the Division were present. Each one was asked to suggest a way out. All types of suggestions were given such as;

- Design and manufacture the gear box in-house
- Try a reputed company in India who are specialists in industrial gear-boxes
- Float global enquiries

I was silently listening but did not utter anything. The MD asked me what I had to say. I told him; can I have a personal audience with you? He closed the meeting and asked me, the GM, and Head of the Fuel Factory to stay back. I said;

Sir,

- None of the suggestions given will work as they are all long-term solutions.

- Please use your good offices and request your friend, the MD of Original Equipment Manufacturer (OEM) abroad (under whose transfer of technology the unit was in production) to send a gear-box from their stock. I further told him that during my last visit to OEM's facility, I was taken around their production and test-facilities where I saw number of such test-stands; so, I am sure they would have a spare gear-box. This would be the quickest.

The MD did exactly that; called his counterpart abroad and **believe me readers**, the gear-box reached us in about a week's time. I was assigned to integrate the gear box in the rig. The rig was made functional and was back in full operation in about four weeks; the fastest recovery plan executed at HAL.

Dear friends: this is the power of 'Relationships'; what huge amount of money couldn't do; one phone call had done.

I have numerous examples like this such as other organizations working exclusively on LCA programme (DRDO-ADA), providing me technical support in respect of my other programmes (non LCA) during their critical phase of execution; with no strings attached. In fact, Relationships and Bonding between HAL and other dealing organizations were the major success factors of the LCA programme.

Therefore, dear friends, before I close on this, let me again emphasise the power of human relationships in execution of a project; believe me, it is tremendous. This I say from my personal experiences. I always put them as number one in my list of priorities; so, go ahead, try and test for yourself; **you will create magic.**

"Treasure your relationships, not your possessions".

-Anthony J. D'Angelo

(c) Technology

Technology is a process which culminates in generating products and services meeting the desired specifications of the customer.

Technology plays a very important role in **'Execution'**. It will be interesting to note that we do not normally fail in **'Technology'** per se, we fail in management and application of **'Technology'**.

There are various aspects of **'Technology'** which we need to understand before taking up **execution** of a project.

Fig. below defines the **'Technology'** chart;

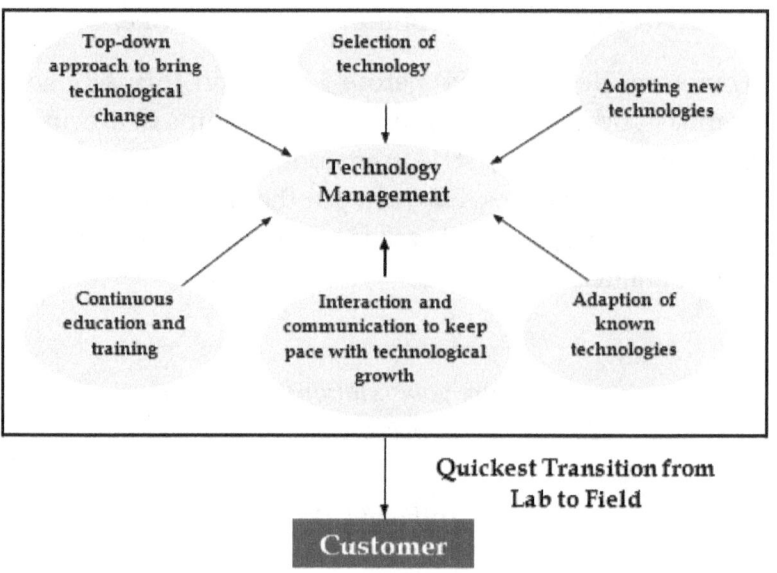

Please remember, successful **'Technology Management'** requires utmost regard to **'Technology'**; therefore, never by-pass **'Technology'**; follow it in letter and spirit.

During my career, I came across number of instances when failures took place and when they were analysed, the root cause was found to be not knowledge but proper application of '**Technology**'. I shall illustrate this with a live example as below;

Example (1980's):

In HAL Lucknow, we had an air-cycle machine called cold air unit used in air- conditioning system of a fighter aircraft. The unit comprised of a compressor and a turbine mounted on the same shaft. It is a compact high-speed turbomachine running at about 65000 rpm. The unit was under transfer of technology from a reputed OEM abroad. When we started first time manufacture of the complete unit, we found we were not getting the temperature drop required from the turbine. The production test-schedule called for minimum of 121 deg.c and units assembled by us were not giving more than 110 deg.c. There were heavy rejections.

The problem was referred to me and my team for detailed investigation. We started examining this. First doubt to be cleared was the test-rig. We took the red-banded unit supplied by the OEM as a standard unit and tested on the rig. We got 136 deg.c which proved that there was something amiss in application of technology prescribed by the OEM during manufacture. We continued further.

Each part was subjected to detailed inspection to see if there were any geometric variations in dimensions. Contours of compressor and turbine wheels and their casings were thoroughly checked. To isolate the problem, we took one set of components from a batch with virtually nil deviations and did matched grinding of compressor and turbine casings with their rotating components (compressor wheel and turbine wheel respectively) and assembled the unit. We tested it and believe me; we got 128 deg.C drop in first attempt. That led us to believe that we should concentrate on tooling which was done. A sensitivity analysis was carried out to identify those features of tooling which affected performance. We found that tooling related to the compressor component is very important for performance of the machine. Some fine tuning of tools used for matched assembly was done. Critical components were manufactured with utmost attention to ensure their conformance to drawings without deviations. Two units were assembled and both passed with temperature drops of 125 to 132 deg. c. There was never a problem thereafter.

This is the power of conformance to Technology.

Here the lesson learnt was that sensitivity analysis of critical components which have bearing on performance and life of a unit is very important. Although in manufacturing environment deviations are bound to happen, this analysis helps in taking engineering decisions.

So friends, always give due regard to Technology. It is particularly relevant to aerospace where every process is well documented. But I think today, all organizations are giving utmost importance to Technology and its Documentation. Therefore, we have to build that culture and never apply short cuts.

"Technology is the skeleton and customers' experience the flesh".

-Shombit Sengupta

Having defined various aspects of Technology, I shall give below a few important points related to Technology.

Innovation

Innovation should become an inseparable part of our **'execution'** process. It is valid across the board; whether it is related to product or process or even organizational structure. There should be attractive incentives in place whereby the innovator is suitably rewarded. An environment is created where innovation can foster. Therefore, it involves all levels of management. It should also be remembered that innovation is not restricted to only a few people; it has to spread across the organization. Finally, please note that innovation does not emanate from suggestion boxes you find in many places in an organization. We have to inculcate a culture of innovative spirit and creativity amongst our work force. A few examples as given below are worth noting:

Organizational Innovation

A big aircraft company abroad which had very tight schedules for delivery to customers found that there was a continuous conflict between shop floor managers, production engineering groups, and designers. They were all located in different places and lot of time was wasted in bringing them together. The Management decided to relocate all of them in one big hall on the first floor of the aircraft hangar where all activities were being executed and a very senior level executive was appointed as coordinator for day-to-day execution. It created wonders.

Customer related Innovation

A company specializing in ATMs found that the press buttons were not of appropriate size particularly for cold countries where users will invariably wear gloves. The gloves had to be removed before pressing the buttons. This was inconvenient for the users. The company realized this and subsequent models had much larger buttons for ease of use by its customers.

Business related Innovation

Take the case of airlines; how the concept of low-cost airlines came into being which has become a universally accepted business model now. It has totally changed the way airlines used to operate. I would call it as one of the biggest innovations in business model of an airline industry.

These are a few examples which illustrate the concept.

Also remember, many times the Innovation is fuelled by the application of Technology; so they invariably go together.

Quality

While we execute a project, we should not lose sight of **'Quality'**. Fast execution does not mean compromise on **'Quality'**. It has to be ensured at all levels of work-flow process. Please remember, **'Quality'** is not only performance, it is all-encompassing and covers even services. Organizations are known by the **'Quality'** standards they maintain; and that is how they build their **'Brand equities.'**

How does an organization build its brand and in common man's language what is the bench mark! It is said that when an organization which is identified by a **noun**; becomes a **'verb'**, that is the pinnacle of its brand value. Take **'Google'** as an example. **Google** as word is a **noun**, but many times you hear people saying for any information they want, why don't you **Google** it! Here, it is being used as a **'verb'**. **That is the brand Google.** Similarly, people have already started saying for taxi service, why don't you **Uber** it. **That is brand Uber.** Similar brand equities have been built by Apple, Microsoft, Amazon, Jio, and so many companies in other sectors such as aerospace, automobiles, pharma, infrastructure build-up etc.

Building a **Quality Brand** is a very strenuous process. It starts from the top and percolates to the lowest bottom; everyone in the ladder of management should be conscious about its value. Remember, it takes years and years to build, and very little time to demolish. We should inculcate a spirit of zero tolerance towards defects that impact **'Quality' and 'Reliability'**.

Training

Training of human resources is very important in **'Execution'**. We have to keep abreast of the technological evolution taking place all the time. For this, it would be necessary to organize well-structured training programmes to cover the complete spectrum of technologies involved in the work-flow process; otherwise, you will be left behind in the race. Each team leader has to have an eagle's eye on what is happening in technologies around the world and see if those can be appropriately adopted or adapted in his/her areas of work.

"Train people well enough so they can leave. Treat them well enough so they don't have to".

-Sir Richard Branson

Founder Virgin Group

B. Execution Made Easy – Operational Level Fig. 3.1 (b)

This is the main part as it deals with the operational aspects of the execution process. Therefore, **Dear Readers,** please focus on paragraphs to follow.

For clarity **Fig. 3.1(b)** has been reproduced as shown.

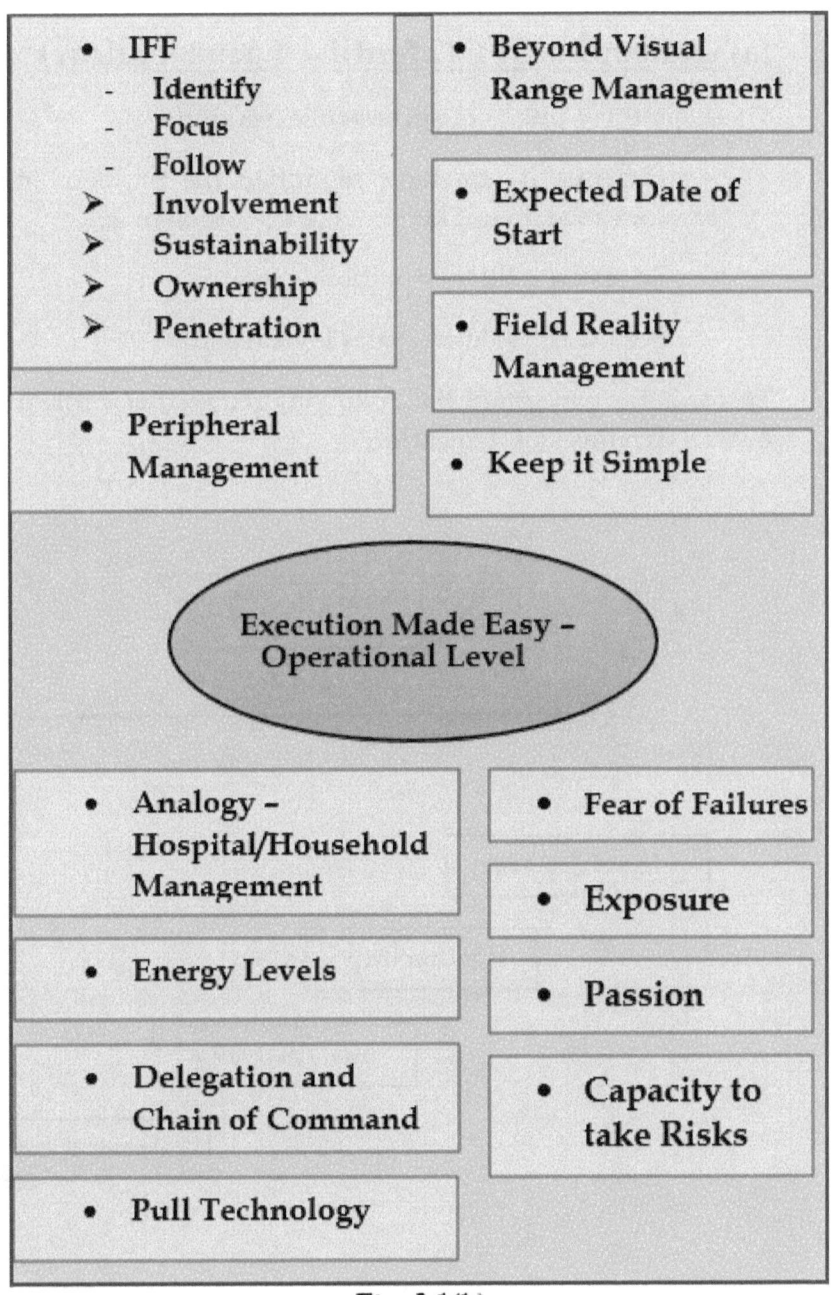

Fig. 3.1(b)

(a) Concept of IFF (Identify, Focus, Follow)

These are three pillars of successful execution.

- First see if you have identified the problem (many times we do not know what the problem is)
- Focus on it till the solution is in place
- Follow the solution to its logical end

Towards the concept of IFF, following play a major role in the overall dynamics of **'Execution'**.

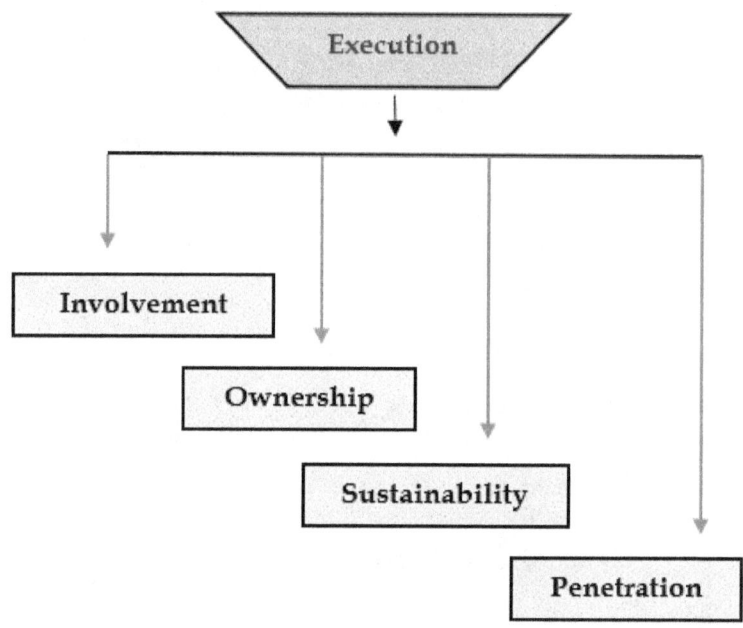

Involvement

Sense of involvement is extremely important. The best from a human being comes out when he has a very high degree of involvement and he takes pride in the job he is doing. He should identify himself with the tasks and involve in his job and his people.

Ownership

Very important. Ask yourself – do I own this project; do I own my resources (capital, human, facilities, etc.)? Do I own these as much as I own my house, my car, my family? If the answer is yes, then half the battle is won. The problem is that generally, people don't feel a sense of ownership. **Can we treat project and its products as animated beings which get hurt if not treated properly?**

Sustainability

'Execution' requires a very high degree of performance on a sustained level; therefore, sustainability is the key. We do not want one-time performers. Many times, failures lead to frustrations impacting the performance of the individual or a group. But here perseverance plays an important role. We need to counsel and motivate our team members so that we get their best performance in a sustained mode.

Penetration

This is an important tool in execution. One has to penetrate deep in the problem to find a solution. Superficially, you as a leader/manger can be thoroughly misguided. There are numerous examples where everything looked bright, but only on surface; once you went deep into it, field reality was vastly different. People are either shy or afraid of speaking out the problems particularly to their boss. So, the leader/manager has to have a high degree of penetration skills to extract the field reality from the team members; it may sometimes look like an interrogation.

I would like to place here an example which appeared in an English daily a long time ago illustrating the concepts described above;

"There was once a King who was very fond of innovative paintings. In his council of ministers and advisers, he therefore, had a chief painter. Once he decided to create a new innovative painting of cocks. He asked his chief painter what should be the criteria to select the best painter. The King was advised that the painting which would attract maximum number of cocks falling, fighting, and swarming over it would be adjudged the best. Painters from all over the kingdom came and made the paintings there which were then placed in a big hall where a number of cocks were brought. Alas! No cock went anywhere near the paintings.

The King was very disappointed that in his kingdom, not even one painter could make this simple painting. He asked his chief painter to attempt one. He said "Your Majesty, I shall take about four weeks to make one". Reluctantly, the King agreed. Four weeks passed. The chief painter came and everyone was surprised to see that he did not bring any painting except some paints and a set of paint brushes in his hands. The King was very annoyed. The chief painter begged for pardon from the King and requested that he be allowed to make the painting then and there. His request was granted. He started the painting and by the time he finished, the cocks (who were again brought in like on the earlier occasion) immediately pounced on the painting."

The King asked; "what were you doing all this time? You could have made this painting during the competition organised four weeks ago". The Chief Painter's reply was;

"Your Majesty! During these four weeks I was not whiling away my time; I spent all my time in the company of the cocks. I slept with them, ate with them, etc. etc. I observed how they behaved, how they ate, how they interacted with each other; I wanted to be natural and realistic to the finish when I make this painting; and the results are evident."

The King and members of his court present were stunned, and the painter was profusely rewarded.

Message for the readers is very clear. If you want to be a true professional, identify yourself with the problem; go deep into it to find a solution; you will create magic. Please remember, **'Execution'** should be based on holistic project management and not on symptomatic management which is what we generally end up doing.

(b) Peripheral Management

While core issues get necessary focus and attention, issues related to peripherals get a back seat and often are the main causes of failure/embarrassment. Typical examples are given in the figure on the next page, shown in the blocks.

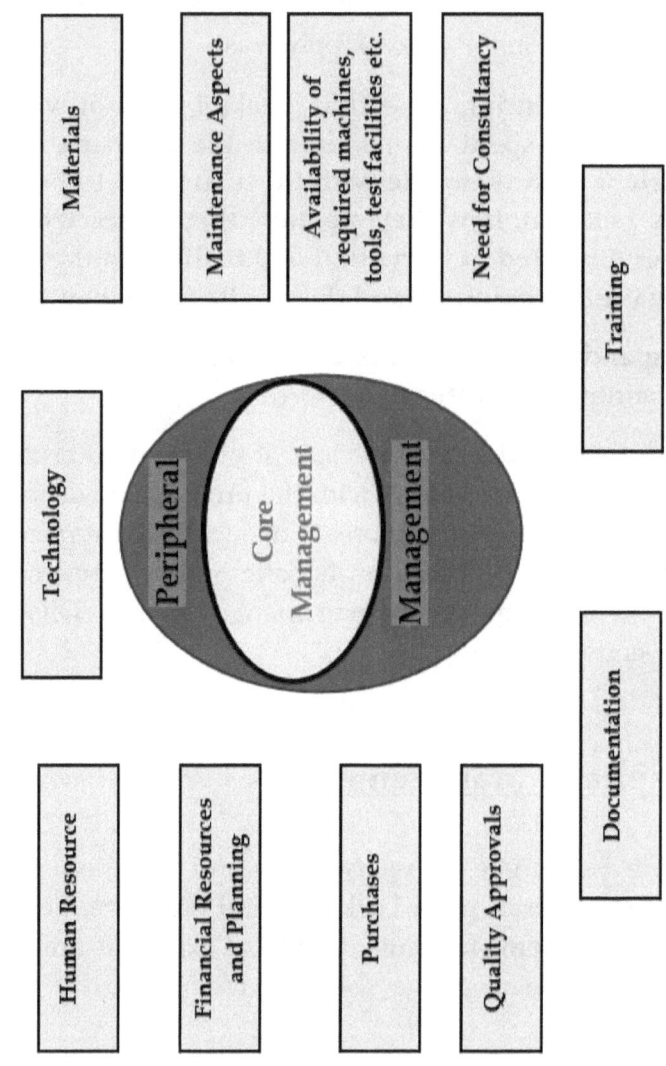

While we follow the core issues vigorously; we should continuously ask ourselves whether the peripherals are under control. If they are, believe me **Dear Readers**, your execution cycle will tremendously improve.

Example:

Procurement of systems and accessories, called Line Replacement Units (LRUs), one of the major peripherals in an aircraft project, can be time consuming and can cause delay if not brought under strict monitoring and control. In our IJT programme, major contribution towards its success was due to in-time delivery of these. To ensure this, I used to conduct half an hour review every morning at 8:30 where attendance of all concerned Department Heads was compulsory. Believe me, that created wonders.

(c) Beyond Visual Range (BVR) Management

This is called **'Anticipatory Management'** which means to anticipate hurdles or road blocks of trivial nature which you as a leader / manager could have anticipated. They are not visible and you have to see beyond your visual range. If you sit down peacefully and write down all that can go wrong, I am sure you can come out with a number of issues which are around the corner but are not visible. Do this in as many details as possible and you will save lot of unproductive time.

(The title BVR is taken from aerospace terminology which means the pilot will deploy the BVR missiles to hit targets which are beyond his visual range)

Here I would like to give a live example. I with my Project Manager of one of the programmes had gone to Air HQs. in Delhi for a presentation to the customer. The presentation did not go well as we were not on schedule and there were a number of technical glitches. I was quite embarrassed.

On the return flight to Bangalore, we both sat down and wrote all the points where we could face problems; some known and some un-known. The list ran into hundreds of them. We also tried to write down possible solutions.

Believe me, these few pages became the guiding documents for us to subsequently monitor, and the result;

"We created an International Benchmark in this project".

This is BVR Management.

(d) Expected Date of Start (EDS)

Dear Readers, I would like to bring-in a new concept here; the **'EDS'**. I am sure all of you will be familiar with **'EDC'**; **Expected Date of Completion**. Normally, this is the date given for completion of an activity.

However, in my experience I found that start of an activity is much more important and relevant. Invariably you go to the shop floor or field and find to you utter surprise that while you have been given the **EDC**, activity has not even started. There could be several reasons for this such as;

- Worker is absent
- Technology not clear
- Material not issued

- Production queries not answered
- Quality Control has not given clearance, etc. etc.

I, therefore, shifted my focus from **'EDC' to 'EDS'**, and accordingly asked my group leaders to ensure that **EDS** for each activity was met, if not, state the reasons thereof, and then focus on them. It made a big difference in the execution of tasks.

While on the subject, let me give another point of view here. I have found that generally workers and technicians want to work because that is why they have come from home; but at the same time, they are rather moody people. They have to have the congenial environment; by this I mean they must have the right implements, right material, clear-cut process and technology, clearance from inspection etc. and anything not in order, switches them off. Imagine an artist; a painter for that matter. If he does not have the right canvas, the right paint, the right brush, best of his creativity does not come out. Same thing applies to the work force also. Now it becomes the duty of shop manager to ensure that everything is in place and environment is just right for every worker to give the best output. I can quote number of occasions when workmen used to approach me asking for more work; particularly the challenging type.

Of course, I admit there may be a few exceptions, but they need to be dealt with separately.

Try this and you will create magic. Your **'EDS' and 'EDC'** both will be by and large met.

In fact, it happens very routinely in our day-to-day life; we keep postponing an activity till we realize it is already too late.

(e) <u>Capacity to take 'Risks'</u>

While executing a project, you as a task-team leader will come across many situations where you have to take **'calculated risks**. Here, your capacity to take **'risks'** becomes very important; and this emanates from the confidence you have, first in yourself and then in your team.

Risk averse managers cannot execute a task properly.

I would like to narrate two live examples here to bring home the point:

Example-1: Rotor Blades of Advance Light Helicopter (ALH) Dhruv

The main rotor blades of ALH made of composite materials were being imported from a firm in Germany. The helicopter was certified with these imported blades. Since this is a strategic item costing a big percentage of the helicopter cost, we were asked to see if these could be indigenized. This was a very **high-risk** proposal considering the fact that main rotor blades are like wing of an aircraft, a **Class-1** item, and if anything happens, helicopter will come crashing down. We took the plunge. This was around 2001 – 2002.

How did we go about!

- Taking reference of whatever drawings and data were available, (there was hardly anything), we analyzed these and carried out complete metrology of the blades.

- Based on this we prepared our own drawings.
- We did a number of tests on material samples to ensure consistency and strength.
- The blades were fabricated to our drawings and then tested on the Whirl Tower and the Ground Test Vehicle (GTV) which were available.
- After having cleared the ground tests, they were integrated on the ALH, and flight-tested.
- After very rigorous flight-testing, the blades were found acceptable and cleared for production use.

From that day onwards, all rotor blades for ALH were manufactured in-house. With over 250 ALHs flying in the field, the total savings in foreign- exchange must have been of the order of Rs. 300 Crore (USD 40 million) at least.

I must mention here that all this could not have been possible but for tremendous support provided by the Test-Pilots, the Certifying Agencies and all the members of my team. The Team was later given an Award by the Ministry of Defence for the excellent work done in Import Substitution.

Example-2: Sea King Helicopters

The electrical wiring of one Sea King helicopter (IN522), which is a part of Tactical Management System (TMS), got burnt during its operation. When approached, the firm, Westland Helicopters UK, from whom these Helicopters were bought in 80s, was asking an astronomical sum to repair, refurbish, and make it flight worthy.

Request was made to HAL around Yr. 2000 to see if HAL could help in its retrieval. We took up the challenge which involved a lot of risk.

How did we go about?

Since wiring was almost completely burnt, we picked up the Tactical Management System (TMS) of another Helicopter (IN523) which was flying, for our study and analysis. We went through the circuit diagrams and also the data available in a few very scanty documents related to TMS and Airframe. With this, we could identify the connectors and other details related to wiring etc. The teams worked very hard and with full support of Indian Navy and Certifying Agencies, the Helicopter was repaired and refurbished and brought to flying status within about eleven months' time; thus, saving valuable foreign exchange to the tune of 1.5 million Pound Sterling at the level of year 2000. Our cost was less than one third.

With the above two examples, see the miracle of taking calculated risks of very high order.

(f) Field reality Management

This is what I call **"Management by walking around"**. What this essentially means is that whatever feedback a team leader gets sitting in office, the situation on ground may be quite the opposite. So, he/she has to find out the ground reality himself/herself which requires him/her to make regular visits to work-centers where the activities are taking place. These are the **nerve centers** of the project. This is because, many times, managers will be given feedback which looks very rosy; they should discard that.

Remember, **'Execution'** has no place for **'arm-chair'** managers who work in **'silos'**.

While on **'silos'**, let me place before you an analogy. So many times, you would have noticed that all concerned departments individually performed very well; but the project could not be executed in time or it failed. Why! It's a problem of co-ordination. Each department should go an extra mile to see as to whatever was advised /suggested/cautioned, has indeed been taken on board or not. This is true even for natural disasters where the co-ordination amongst various agencies such as Meteorological Department, State and Central Govt. Agencies, Disaster Management Team etc. is very important; otherwise, disaster is just around the corner. I am putting so much of emphasis on this because of my experience in aerospace where without intense co-ordination by the team leader/project manager, the project will not move fast as it involves multiple disciplines, departments, and organizations. In fact, to my mind, it should be equally true in the medical field where new advancements in medical science have led to a strong need for overall co-ordination.

Here, I would also like to narrate my experience when I along with my Chief Designer, visited Dunlop company in Coventry England in mid 80s. We had Design Collaboration with Dunlop for Aircraft Wheels and Brakes.

We made a courtesy call on the Managing Director of Dunlop (worldwide) and had discussions on the modalities of our working together. After the meeting, on our way out, we found an ante room which was his office. Out of curiosity we asked him how he was managing worldwide operations with such a small office. He spoke and I quote; **"I don't need it; all I want is my Personal Assistant (PA), and a young and energetic Co-ordinator. That's all"**. Unquote. He had full knowledge of what was happening in all divisions of Dunlop all over the world. Imagine this when there was no internet and communication system also was very poor.

Dear Readers, I place so much importance to overall co-ordination that I strongly feel there should be a separate **Ministry of Co-ordination** in both, State as well as Central Administration to co-ordinate and give continuous feedback to the Head of the Govt. I am sure it will make a huge difference in governmental administration.

(g) Keep it Simple (KIS) Concept

Here I want to quote Michael Dell, founder of DELL Computers who used to say, **"whatever is easy to operate, will sell more"**. So, remember, latest technology does not mean a complicated product. For its easy operationalization in the field, **keep it simple**.

(h) Analogy

With Hospital Management

I want to picture a scenario here and then draw a comparison. You would have come across a number of instances where an activity is ready to start but is waiting for the covering documents. This is very common in manufacturing environment in a factory.

Imagine, does this happen in medical profession! Even in an old hospital in a remote area, the prescribed system works and all documents related to the patient accompany him wherever he is taken for procedure. Why can't we emulate what our medical brethren are doing! It is a question of casual attitude and mindset. We have to change this to cut down non-productive time. Any time wasted is after all money.

Time has to be accorded utmost respect.

With Household Management

I am sure all of you will agree with this analogy. We have to learn from the housewife how she manages the kitchen; the supply chain, quality of food items, the ingredients used therein, the storage of food items in right conditions, inventory control, replenishment, and so on and so forth. You would have never faced a situation where you reached home tired, and food was not available. Even if you bring some guests with you un-announced, she will somehow manage. She is also apt in multitasking and financial management. If our housewives can do, why can't we with all the support that we have? It's a question of **'involvement'** and **'focus'**.

---

*"I will love the light for it shows me the way;
yet I will endure the darkness because it
shows me the stars"*

-Og Mandino

(i) Fear of Failures

Many times, we get disheartened when a failure takes place. I have an altogether different view here. I think **'failures'** are indicative of **'progress'**. More 'failures' mean, we indeed are progressing. Please remember the following;

- Success comes from experience and experience comes from **'failures'**.
- It is not important why you fell down; more important is why you slipped.

We should widely publicize failures for people to learn lessons. However, we must remember one important aspect of **'failures'**; there is no mourning period in the process of **'execution'**; show must go on. Let me explain this with a live example;

In one of our trainer aircraft development programme, during testing on ground, a large number of push-pull rods which operate the control surfaces, failed when design loads were applied. This happened when we were just **ten weeks** away from our scheduled and already committed maiden flight in first week of March 2003. In aeronautical parlance, this will be treated as a major failure and hence a major set-back. How did we recover! See below;

- Had thorough investigation done with specialists in the field.
- Informed the Certifying Agencies (very important).
- New design carried out within two days.
- Bought out items (bearings, seals etc.) procured by paying extra cost for accelerated deliveries.

- Manufacturing as per new design initiated in three wok-centers concurrently, two in Bangalore, one outside. One **co-ordinator** was positioned at each of these work-centers.
- Assembly and re-test done, and items delivered to aircraft after clearance by Quality Control and Certifying Agencies.

All this was done within six weeks from the incident. Everybody worked on war footing, and the result;

The aircraft made its maiden flight on the scheduled date of 07 March 2003.

A point to remember here is that capacity of all of us goes up many folds when you are in crisis and that is the one which comes into play when we do not look back and move on. Please remember;

"The greatest glory in living lies not in never falling, but in rising every time we fall".

-Ralph Waldo Emerson

(j) Energy Levels

Energy levels in the task team are very important. The team should be very dynamic in nature at all levels and at no point of time should appear to be static without any direction. There are many factors which bring dynamism such as;

- leadership
- well defined tasks
- role clarity
- aligned vision towards goals and objectives
- working environment
- motivation

The project team should realize this and bring dynamism in its operation; something like spirit of **'restlessness'**. The team members should work on a campaign mode on the pattern of a **Control and Command Room of Armed Forces.** They should also be ready to change for the better since the outside world today is changing very fast as to how the business is to be conducted. Please remember;

"When the rate of change outside exceeds the rate of change inside, the end is in sight".

-Jack Welch

Every team member should be prepared to endure pain associated with change and should not become victim of status quo.

(k) Exposure

Exposure of your members to top level in the organization is very important. This will bring lot of confidence in them and they will get motivated to give their best.

When I joined Aircraft R&D Center of HAL at Bangalore in Oct.'95, I found to my surprise that even senior designers/managers in my team were of shy nature. They would not like to be exposed to top management of HAL to the extent that if there would be a visit from Chairman/Directors of HAL or other VVIPs/VIPs to our facility, they would not come in the front and would like to stay at the back. Since we were working on LCA project which had very high visibility, these visits were very frequent; both from India and abroad. I thought this was not good for their growth and advancement in the organization. Therefore, I started exposing them, bringing them to the front; and see the advantage over a period; they and their work became known to the top management of HAL and this helped them a great deal in advancement of their careers. In fact, I can particularly quote few cases who rose to the positions of Director and General Manager in HAL.

This is advantage '**Exposure**'.

(l) Passion

Very important. You should be very passionate about your profession and whatever you are doing.

In this context I would like to recall an incident when I was at Honeywell Normalair Garrett Ltd, UK from 1972 to 1974 undergoing a two-year post-graduate design course. At the start of the course, I had to undergo a one-week training on mechanical drawing and drafting under the Chief Draftsman of the Design House.

On the first day, he gave me the original of a very complicated front view drawing of a compressor disk with integral blades for my study asking me to come back with doubts. I must confess that the drawing was very neatly done; it was like work of art. I started going through and with great difficulty could locate the center point from where all references for drawing of blades were taken. To facilitate future study, I put a very small circle of say 1 mm dia. around that center point. The Chief Draftsman came back and when he saw this, he became very upset as to why I spoiled his drawing. In spite of my tendering an apology, he stopped my training for two days, and only after that he started my training. This is how the passion works. He told me later that while all his creations were nothing but drawings, he treated them as his children; and anybody putting his finger on them would hurt him no end.

Can we bring this level of passion in our work? We tried this in our projects. In fact, we made one senior designer/manager as vehicle coordinator (VC) for each prototype aircraft. This really worked. The VC took total ownership of that vehicle and became so passionate that any shortcoming of that vehicle would hurt him/her no end. The VC would go out of the way to address that so as to bring the vehicle back into the flight stream. There was a healthy competition amongst vehicle coordinators (VCs) as to whose vehicle would fly first.

And result; **a noticeable improvement in our flying rate.**

(m) <u>Delegation and Chain of Command</u>

While a leader/manager builds a team, he/she has to consider two very important aspects of execution, **Delegation and Chain of Command.**

Therefore, the team leader has to delegate to levels below with well-defined roles and responsibilities. Many of our leaders and managers feel reluctant to delegate and try to handle everything themselves. This is no way to build the team. A leader/manager should realize that if he/she starts doing their work, who will do his/her work. So, my advice, please go-ahead and delegate as much as you can with overall perspective in mind. Keep an eagle's eye on how your team members are performing and make mid-course corrections where required. Remember, responsibility will still be yours both, in technical matters as well as in commercial/financial matters as you are the leader. Also keep this in mind that there can be only single chain of command; no parallel power-centers; otherwise, whole system will collapse.

In this regard, the example of our Armed Forces is very apt as they follow unified chain of command. Let us emulate their practices in management of our projects. But like in Armed Forces, here also the leader/manager has to lead from the front.

(n) <u>Pull Technology</u>

This is something new which I want to talk about. **Pull Technology** is an important concept for effective workflow process for each of the activities. It can be explained through the diagram shown on the next page.

Pull-Technology Chart

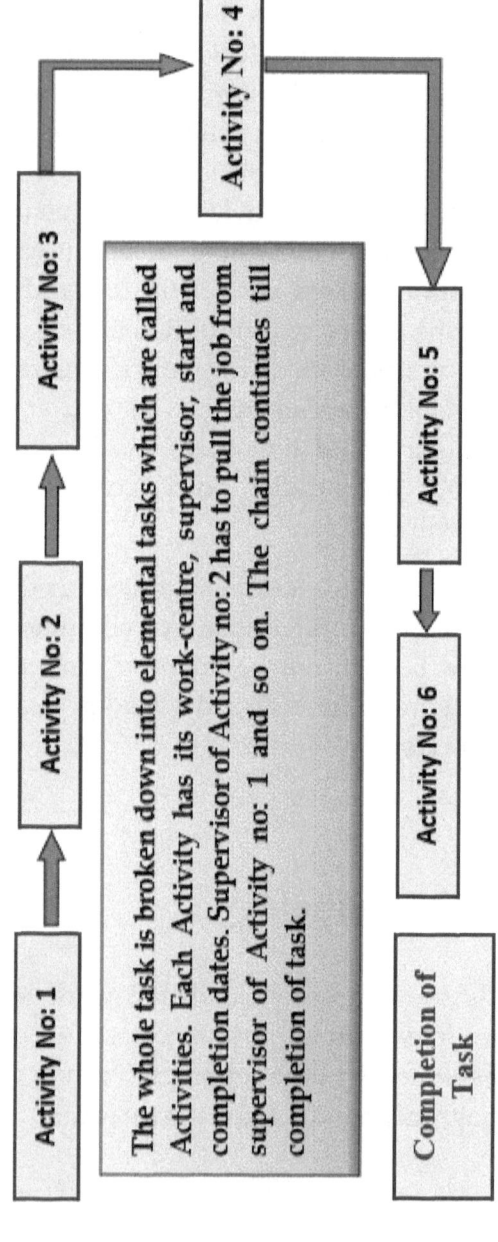

Activity No: 1 → Activity No: 2 → Activity No: 3 → Activity No: 4 → Activity No: 5 → Activity No: 6 → Completion of Task

The whole task is broken down into elemental tasks which are called Activities. Each Activity has its work-centre, supervisor, start and completion dates. Supervisor of Activity no: 2 has to pull the job from supervisor of Activity no: 1 and so on. The chain continues till completion of task.

Pull technology concept is explained in the following steps;

- Overall task is broken down into elemental tasks called Activities.

- Each Activity has an assigned work-center, a supervisor, start date, and completion date.

- Activity no: 1 will flow into Activity no: 2 and so on, till completion of Task.

- Each work-center will define its internal and external customer; for instance, work-center of Activity no:1 is internal customer of Activity no: 2 whose external customer is work-center of Activity no: 3.

- Each recipient work-center has to pull the job from its internal customer, process the activity, and deliver to its external customer; and this goes on till the task is completed. The supervisor of each work-center has to be agile to pull the job from previous work-center and should do all preparations in advance to process and deliver to the next work-center.

The success of this technology depends on a few parameters as below;

- Breaking the task into elemental activities.

- Agility of each supervisor.

- Preparedness of each work-center. Once I know that the job is coming to my work-center, I should be ready with all implements in place; not that I start looking for them after the job has arrived; a practice you will find very common. In this context, I want to bring-in the concept of Indian wedding. **When the marriage party and groom are at your doorstep, you don't go looking for garlands.**

- The whole concept here is to reduce un-productive time to minimum and bring more efficiency in the work-flow process. Try this; inculcate this culture in your team members; you will find the difference.

With this we finish the operational part.

C. Execution Made Easy–Review, Monitoring And Control – Fig. 3.1 (c): Review Mechanism, Project Manager

(a) Review Mechanism

An effective **'Review mechanism'** is the most important tool of **'Execution'**. However, some of the essentials of a **'Review mechanism'** are;

- It has to be short
- It has to be dynamic
- It has to be regular

I am of the firm view that all projects, but definitely the ones which are put on the fast track and are high profile, need to be reviewed with **top-down approach**. Now, since neither the top executive nor the team leader/manager has so much of time at his/her command, a very simple and yet effective **'Review Mechanism'** proforma was evolved first time by me which I illustrate below with a simple diagram **(Fig.3.2)**.

The core activity; in this example Flight Clearance of Aircraft, is placed at the center of the Ellipse, while sub-activities responsible for the completion of core activity are placed around it. One can put start and finish dates also for each block of sub-activities. Then each sub-activity can have its own Ellipse and so on.

With few such Ellipses defining the core-activity, the whole review can be completed in about an hour every day; enough to monitor the project on a daily basis and put it on fast track.

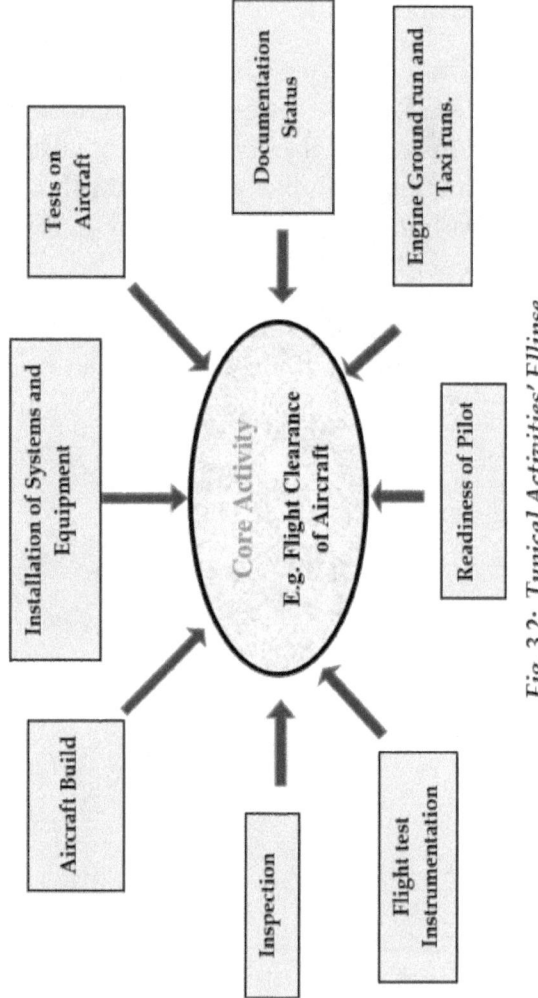

Fig. 3.2: Typical Activities' Ellipse

Believe me **Dear Readers**, I used this very successfully in my day-to-day management of the **Light Combat Aircraft (LCA) 'Tejas' programme**. Just about ten such **Ellipses** were enough for a grip on the programme on a daily basis. The benefit of this quick review mechanism was such that I subsequently started using it for all major programmes.

In this context, I would like to bring one point very clearly. This technique is not to replace the classic PERT and Bar charts which will continue to be used for overall review; may be once a month. This Review Mechanism is for daily and quick review of the project.

I have named this as YK* Activities' Ellipse Review Mechanism

(*Yogesh Kumar)

In addition to this, for some critical activities, it may be desirable to form;

- Crack teams
- Rapid action teams;

depending upon their criticality.

The fundamental to all above is the concept of **'elemental build-up'** of the task. If this can be achieved, more than half the battle is won. Every task looks big to start with but when broken down in small tasks, it looks simpler. This is the core concept of the **'Ellipse Review Mechanism'** illustrated above.

Many times, while reviewing a project, we tend to get lost in smaller issues and as a result lose sight of the bigger picture. Please do not fall into this trap. Start from the bigger picture and then go down the line; you will definitely succeed. Also please remember, last ten percent requires almost ninety percent of effort.

(b) Project Manager

The concept of **'Project Manager'** for the programme is very essential to co-ordinate various activities of the project. Broad roles and responsibilities of the Project Manager are;

- To produce the specific end product within the technical specifications, cost and time schedules, with the available resources.

- Alert the management at the right time (watch word **'alert'**).

- To make or force right decisions at right time (watch word **'force'**).

- To interact with other Managers.

- Prime single point contact.

- Conflict and crisis management.

- To **foresee** the unpredictable.

It is generally seen that in the initial stages of the project, the **Project Manager** at times, gets into conflict with his/her counterparts particularly from support services. (Finance, Commercial, Maintenance etc.)

While exhibiting dynamism, he/she should show a lot of patience, balance of mind, and perseverance. Everything settles down in the end. I always used to say that;

- More the conflict, more the progress
- An unpopular Project Manager is generally a successful Project Manager

It will be appreciated that Project Manager has a difficult task, and in the execution of the same which he/she does with no holds barred, he/she may antagonize people as he/she moves along.

Therefore, needless to say, he/she has to have blessings of the Team Leader or Head of the organization who should provide full support to him/her. My suggestion is that the Project Manager should be high-up in the hierarchy of the organization.

However, before I close on Project Managers, I have a suggestion. To implement mega projects on infrastructure development, our country will need **Project-Managers** with eye on **execution** in large numbers. I strongly feel that '**Project Management'** with major focus on **'Execution'** is a discipline which should get due recognition and attention in our Academic institutes, R&D organizations and Industries. A short exposure to this would help our engineers and managers to take up the challenges of **'Execution'**.

Of late the **'finishing school'** concept is gathering momentum. In this, fresh engineers are inducted by the industry and then subjected to a much-focused training programmes which cover both, the technical side as well as the management side. This is to make them appropriately useful to the requirements of the industry. May be during this course, a capsule on **'Execution'** with **Project-mode Management** should be included. It should be based on a large number of case studies presented by eminent personalities who can share their practical experiences.

"Great works are performed not by strength, but perseverance".

-Samuel Johnson

Dear Readers, in the above paragraphs, I have brought before you some very useful tips on '**Execution**' based on my personal experiences. I have tried to capture very wide spectrum and **I can guarantee that if you follow these, you will surely be a winner.**

So, go ahead and try for yourself, success will be yours.

However, before I close, I would like to bring out one important aspect of '**Execution**' which is related to the organization responsible for the task(s). You would appreciate that for the **Chief Executive or the Task-Team Leader** to deliver the best, commensurate support from the organization is also vital. At no point of time, the organization should be found wanting. Towards this, I have thought of a few organizational points which to my mind are vital and merit consideration.

Organizational points

> *"The rise of nations comes with an increase of men of character and of strong ethical and moral fiber".*
>
> *-Swami Vivekananda*

(a) Motivation

The work and project in themselves are a great source of **motivation**. You cannot imagine how intoxicating the thrill of working on a project can be, provided there is a conducive working environment and a leader who cares. If this was not the case, I don't think our **Team-LCA** would have worked voluntarily almost **twelve hours per day**, for all the six days a week, for eleven long years. Sometimes, I had to virtually force them to stop work and go home. Therefore, I am more than convinced that **real motivation flows from the power of the projects.**

Our motto during execution of projects was; "**whatever we do, nobody else should be able to do better**". We have to overcome the mindset that we cannot do it.

Therefore, the organization should ensure continuity of projects to keep the work force motivated.

Let me give an example here. There was a long gap of about fifteen to twenty years between the last design project **'Ajeet'** taken up in early seventies and the **LCA** launched in eighties/early nineties. This much of gap can be disastrous for a Design Organization as it gets totally demotivated as was evidenced in the beginning of the LCA programme, where we started with hardly any infrastructure by way of trained and skilled manpower, computational facilities, technology infusion, plant and machinery, test-facilities etc. We had to build all these step-by-step concurrent to the development programme; a great challenge in project execution.

Towards this, I believe that each organization should have a **'Vision'** document in accordance with its **'Mission'** statement which lays down the route it will take for its growth in the next say ten years. That would be the guiding document for taking up new projects and develop commensurate infrastructure.

Another important point I want to make here is that most of the major R&D and Infrastructure development programmes are likely to be in Govt./Public Sector till a viable system is in place towards which efforts are on. Although it is true that this imposes certain limitations by way of inflexibilities, but the freedom a leader gets in execution of the projects, more than offsets this. The dealing organization should provide an enabling platform to its leader so that he/she **brings private-sector efficiency in Government /Public Sector environment. (e.g. in LCA).**

(b) Top down Approach

Top-down approach is very important in any project, particularly the one which has a large innovative content and is multi-disciplinary and multi-organizational. This requires total involvement of the top management team. The message is driven very clear and everyone gets focused and involved. Therefore, the organization should inculcate that culture. I followed this practice myself in all the programmes I was leading and it made lot of difference. I told my top and senior management teams also to follow this approach.

(c) Decision Making

An organization should be proactive in decision making. A right decision taken late is as good as taking no decision at all. This is particularly true for high technology, high stake projects where timing is very important. Top-down approach suggested above will, to a large extent, facilitate this.

(d) Operationalization

A task is complete only when it is fully operationalized; that is customer has started operating it. Therefore, issues such as customer orientation, maintenance, and after sales support etc. have to be addressed by the organization from the very beginning in line with the technology. That is where **value engineering** assumes a significant role with initiatives like design for manufacturing, design for inspection, and design for maintenance. It covers the full life-cycle of a project.

(e) Localization of Management

Managing a team is largely dependent on environment and people. What will hold good in production environment, may not hold good in an R&D environment or vice-versa. Also, the work culture may be different in different geographical locations of the work-centers. The organization should be sensitive to this and formulate its policies regarding human-resource management accordingly.

(f) Grain Boundaries

Many of the projects in today's world are multi- disciplinary and multi-organizational. The task team leaders therefore, have to deal with different organizations. Their parent organization should create an enabling environment and facilitate them to dismantle/remove these so-called boundaries as problems are always in the interfaces. I call them as grain boundaries; a terminology taken from metallurgical engineering related to aircraft gas turbine engines. In an engine, the turbine blades which see the highest pressure and temperature have to be of highest strength. To provide that strength, the latest technology is to use single crystal blades with no grain boundaries. The same concept has been brought here.

Please remember, relations between dealing organizations can make or mar the project.

(g) Product Engineering

This is a discipline on which lot more emphasis and focus are required on the part of Academia, R&D Organizations, and Industries. It has to get the importance it deserves. Eventually, every effort or a task culminates in a product or service or both. We have seen that those organizations which focused on this created magic; both abroad and in India. Imagine what Steve Jobs of Apple used to say **"don't ask customer what he wants, tell him what he needs"**, and rest is history. Today prospective buyers stand in queue for hours together to buy its products when they are launched first time.

On a similar count, take the case of automobile scenario. The best design of automobiles comes out from extreme focus on product engineering and management; be it aerodynamics, ergonomics, metallurgical engineering, engine design, comfort considerations, safety, reliability, and so on. Every year new models are launched in the market. They are displayed in the reputed expositions world over and their features explained. Their motto also is telling the customer what he needs.

Therefore, a growth-oriented organization has to put lot more focus on product design. But remember, **best product designs are the ones which are evolved on the shop floor**. Therefore, organization should evolve a method of receiving inputs even from workers and technicians to fine-tune its products, and their related services. I did it very comfortably during my career and believe me, I always benefitted and honed my technical skills.

All the above points related to organization are shown in a summary form at **Fig. 3.3 as ready reckoner.**

Dear Readers, in this chapter which is centering around the main theme of the book i.e., **'Execution'**, I have placed before you practical tips which in my experience are very vital for successful and fast **'execution'** of a project. I have also brought out a few vital points related to the dealing organization.

I have no doubt in my mind that they will be immensely beneficial to both, the Task–Team Leaders and their Organizations.

"Men often become what they believe themselves to be. If I believe I cannot do something, it makes me incapable of doing it. But when I believe I can, then I acquire the ability to do it, even if I did not have it in the beginning".

-Mahatma Gandhi

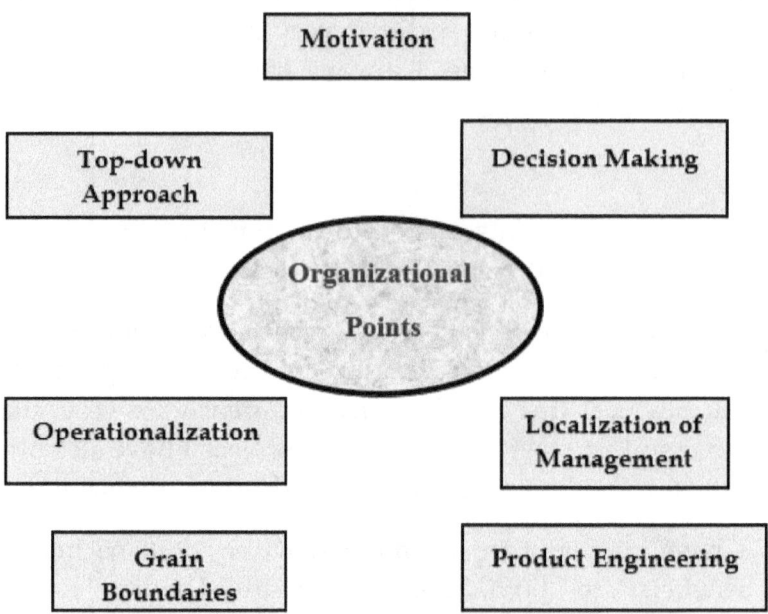

Fig. 3.3

CHAPTER FOUR

And the Results!

> *"Aeronautics was neither an industry nor a science. It was a miracle."*
>
> -Igor Sikorsky

Dear Readers,

With so many practical tips I gave you in the previous chapter, you might as well be asking in your mind, **what were the end results!** Did we meet our aims and objectives?

Well, let me tell you very humbly, that me and my team put – in their best efforts and the results were forthcoming. The summary and the pictures that follow bear testimony to this.

Light Combat Aircraft (LCA)-Tejas

The team LCA of about 1000 engineers, designers, workmen, and technicians put an average of seventy hours per week for eleven long years and the results;

LCA – Tejas is now in regular production at two Divisions of HAL. It has the highest indigenous content amongst all other programmes including ones under transfer of technology and has about seventy percent work share of private sector companies. With great pride, I am happy to inform that the aircraft has already been inducted as a front-line fighter in the Indian Air Force; a proud moment for any aircraft programme.

Sight to see so many LCAs in the Squadron of Indian Air Force

Armed to the Teeth

LCA was one programme which brought so many diverse organizations and people together with one aim and one goal. It therefore, set a unique example of this phenomenon **which perhaps does not occur very often in history.**

While on the subject, it will not be out of context to state that LCA programme has also been the most criticized and written about programme in the history of military aviation in our country. But, as said earlier, the objective was not only to build aircraft but also develop key technologies and set-up facilities concurrently for this and the follow-on programmes. This has actually happened as there are a number of follow-on programmes which both, DRDO-ADA and HAL are working on, including the future variants of LCA.

Further, development of such a complex vehicle takes nothing short of fifteen to twenty years from concept to operationalization even in the most advanced countries which have a legacy of such programmes.

The success of LCA is the best example of Perseverance, Hard work, Synergies, Management of Technology, and **Project-mode management.**

What has LCA Programme Achieved!

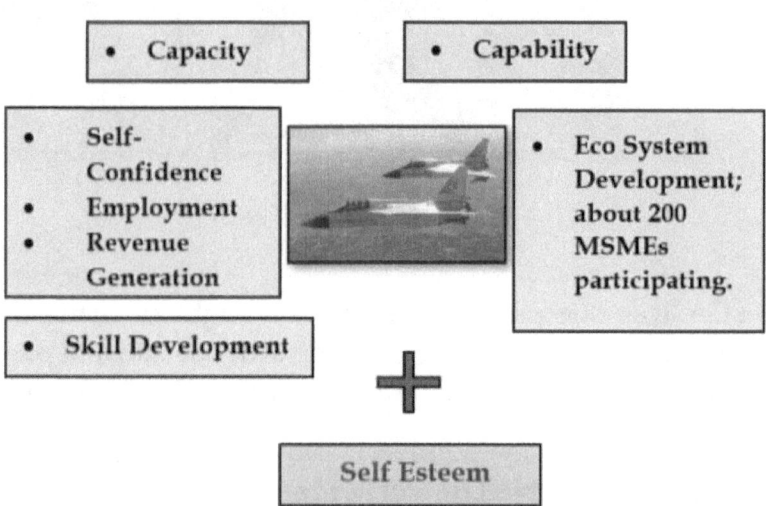

- Capacity
- Capability
- Self-Confidence
- Employment
- Revenue Generation
- Skill Development
- Eco System Development; about 200 MSMEs participating.

Self Esteem

Dear readers, before I close on LCA, I shall make one point that LCA was the most rewarding experience of my career. I have always said, development of LCA cannot be treated merely as development of a product or technologies. The whole programme was like a **'movement'**; it was to be the manifestation of our dream of becoming self-reliant in this very vital and strategic field of military technology; and we succeeded in achieving that.

I am convinced that LCA with its path-breaking technologies is a major step towards self-reliant India, (Atmanirbhar Bharat).

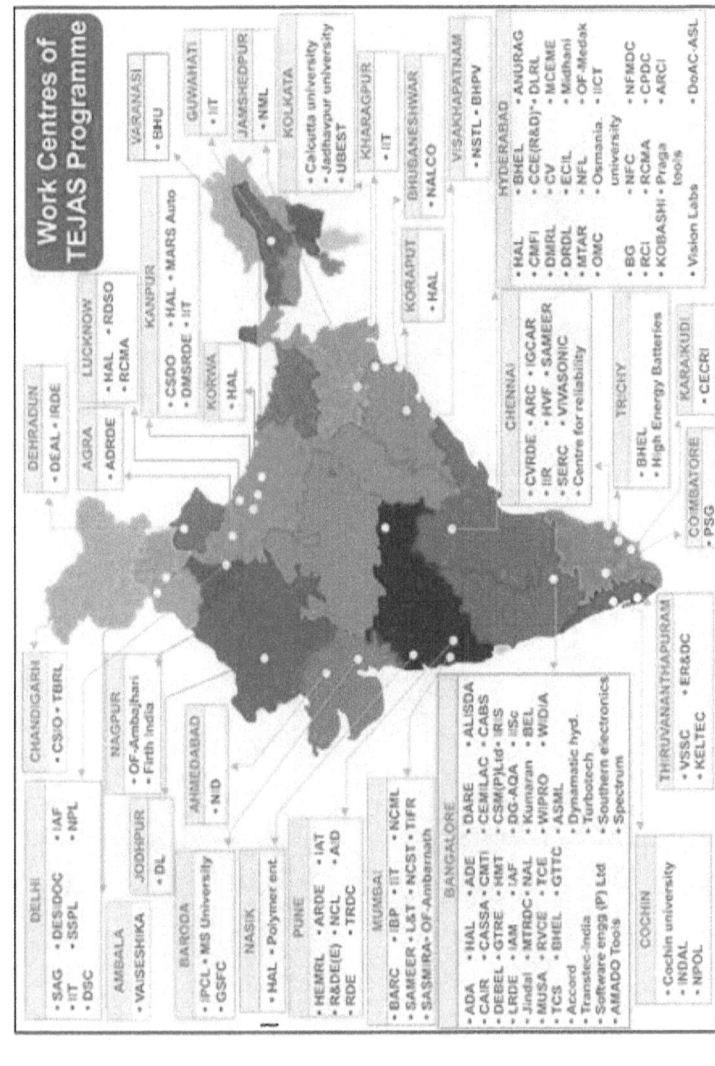

It can be seen that the whole country was associated in the development of LCA – Tejas; (Major challenge in Project management)

Fortune favours the brave- Virgil, Aeneid

A historical milestone in Military Aviation

Maiden flight of the LCA Technology Demonstrator (TD-1) on 4 January 2001

Intermediate Jet Trainer (IJT)

- Overall development cycle time reduced by an average of **forty percent**; possible only through digital design and implementation of Product Life Cycle Management (PLM).
- Time to flight reduced by about **twenty percent**.
- Cost of acquisition reduced by about **fifteen percent**.
- Technology transfer to manufacturing division through digital route.

Milestones Achieved

One of the criteria in aircraft development programme is the time taken from start of manufacturing (first metal cut) to first flight. Normally, this is around twenty-four to thirty-three months depending on complexity of the aircraft; we achieved this in **twenty months**. Moreover, this was achieved in two consecutive prototypes PT-1 and PT-2 as shown in the following table;

Table: Major Milestones-Start Dates

Sl. No:	Activities	PT-1	PT-2	Months from Metal-cut	
				PT-1	PT-2
1.	Metal cut	Jun 01	JUL 02	-	-
2.	Assembly & testing	Jan 02	JAN 03	7	6
3.	Maiden flight	7 MAR 03	26 MAR 04	20	20

With the above milestones, we created an International Benchmark in this project

Award Won

The Team- IJT won the **"IPMA International Project Management Award"**, (considered an **'Oscar'** for excellent project performance) in the year 2005 for its Project-mode Management approach in this programme which put it in top league.

Paris Experience

The aircraft was taken to International air show in Paris, the most prestigious show in the world in June 2005. It flew everyday dot on the appointed slot and came out with flying colours.

It is difficult to express in words the excitement it generated amongst the delegates who had assembled from all over the world.

Solid Modelling

Decision to Adopt 3D Modelling Technique for Design; largely responsible for reduction in development cycle as mentioned earlier.

Chief Test Pilot Sqn. Ldr Baldev Singh of HAL (in uniform) inspecting the first IJT prototype before its maiden flight on 7 Mar. 2003

"Proof of the Aircraft is in its flying"

First Flight of IJT- 7 Mar. '03

First Flight of 2nd Prototype on 26 March 2004

Two Prototypes flying together

The IJT on display at Paris Air Show, 2005. Left to right: Mr. Jayasimha, Chief Designer & Project Manager, Mr. Yogesh Kumar, Executive Director, ARDC, Mr. Ashok Baweja, Chairman HAL, Mr. Shekhar Dutt, Secretary Defence Production & Mr. Pushpindar singh

Jaguar Avionics' Upgrade – DARIN - II

A unique example of concurrent design and production; just look at the results in Table below;

Table – Milestones Achieved

First flight	29 Sept. '03 (12 months from start in Sept.'02)
Initial Operational Clearance (IOC)	29 Sept. '05 (24 months after first flight)
Final Operational Clearance (FOC)- Trainer	15 Sept. '06 (36 months after first flight)
Production aircraft delivery- Trainers	Eight – 31 March '04 Nine – 31 March '05

> Above milestones broke all records in HAL and created International Benchmarks.
> The Flight-Test Team conducted record Sixty test-flights in just two months on production aircraft.

Operationalization

This is one project which has been fully operationalized with the customer, the Indian Air Force. During the last so many years, feedback has been very good and all the pilots who have flown are upbeat about the system and its capabilities.

Spin – offs

The experience gained in this programme led to a number of important follow-on programmes at HAL such as further upgrade of Jaguar, Mirage and other aircraft.

Award Won

Excellent effort put-in by the Team-Jaguar was duly acknowledged by the Ministry of Defence and the team won **"Raksha Mantri's Award for excellence; 2003-04.**

"For Import Substitution"

Nothing can be a better recognition than this!

Jaguar Cockpits	
Pre – mod (1982)	Post – mod (2002)

Production Jaguars (Trainers) ready for delivery to the IAF in July 2005; less than three years from start of the programme with all modifications complied and certified, setting an International Benchmark

Raksha Mantri's Award for Excellence

2003-2004

"For Import Substitution"

—————— Awarded to ——————

S/Shri Yogesh Kumar, P.L. Vaishampayan, S.P. Bhattacharya and Y. Kumar of Aircraft Research & Design Centre Division of Hindustan Aeronautics Limited, Bangalore for sustained indigenous development efforts, quality conscious approach, pursuance of excellence and taking initiative for achieving foreign exchange savings through development of Software for Mission Computer.

★

K.P. Singh
Secretary, Defence Production

Pranab Mukherjee
Raksha Mantri

New Delhi
June, 2006

Shri Yogesh Kumar, Director (Light Combat Aircraft), Hindustan Aeronautics Limited, receiving Raksha Mantri's Award for Excellence 2003-2004 in Import Substitution Category for Development of Mission Computer Software on 9th June 2006 at New Delhi

Revenue Generation

HAL being an Industry, every project is looked at from the revenue it has generated. I am proud to say that all the three projects I mentioned in the beginning, have very high potential; the top of the chart is of course **LCA – 'Tejas'** as shown below;

LCA 'Tejas'	: Rs. 50000/- Cr (6 to 7 BUSD)
IJT	: Rs. 9000/- Cr (1.2 BUSD)
Jaguar Avionics' upgrade DARIN – II*	: Rs. 1700/- sales turnover in two years (Much more to accrue from fleet modification)

The above figures are on conservative side; actual potential will be an order of magnitude higher over a period of time.

---

*A rough estimate indicates that there was a saving of about Rs: 800 Cr (USD 110 Mi) at 2001–2002 level of economy in the Jaguar programme alone. More importantly, it gave the confidence to our designers and engineers for future programmes.

Dear Readers,

The decade starting 2001 was an important era in aeronautics when three major programmes took shape in fixed wing design, the LCA, IJT, and Jaguar upgrade. They not only motivated the entire aeronautical fraternity, but also gave different and definite directions to various organizations.

CONCLUSION

> *"Things do not happen. Things are made to happen".*
>
> *-John F. Kennedy*

Dear Readers,

In the previous Chapter, I have placed before you three distinct examples, each very unique in itself. While **LCA** was a challenge in technology and project management, **IJT** was in effective and fast **'execution'**, and **Jaguar** in concurrent design and production. All the three extensively used the principles of **'Execution'** stated in Chapter-Three.

I have always held a view, which I have stated earlier in the beginning of the book also, that we fail to deliver not because of lack of knowledge; it is only that we are not able to **execute** the project well enough.

In today's world, projects are becoming more multi-disciplinary and multi-organizational, with lot of deployment of technologies. Therefore, **'execution'** is becoming more and more challenging.

Other sectors

A few thought processes placed before the esteemed readers in the book, although derived from experience in aerospace sector, but they are equally, if not more, valid to any sector of our economy; whether it is the Railways, Road building,

Infrastructure development, Hospital management, or any sector which calls for managing large programs. I would even go to the extent of saying that they are equally valid in day-to-day administration and governance.

Summary

In the end, I would summarize as below;

- Nothing is impossible provided we focus in the right direction with aligned vision.
- We have a very strong talent base in our country with our younger generation full of energy and bright ideas; all we need to do is to channelize them towards a common goal.
- Our country is in take-off stage and has to move fast towards path of development to become a fully developed country.

So, plunge deep, follow the practical tips on 'Execution', give utmost importance to the esteemed customer from the very beginning, address adequately the issue of concept to delivery, and I am sure, success will be yours.

Therefore, think big, act big, and achieve big with speed and scale, and set an example for others.

"If you want to shine like sun, you have to burn like sun".

-Dr. APJ Abdul Kalam

ACKNOWLEDGEMENTS

First of all, it is my proud privilege to express my sincere gratitude to my parent organization, the Hindustan Aeronautics Limited (HAL), which gave me tremendous opportunities towards growth of my career from young management trainee in 1969 to Director (LCA) in 2006 (**Board level position**). During this period of about four decades all through in Research, Design, and Development, I had the rare privilege to work under the guidance of larger-than-life personalities, both within HAL and outside, who made a difference in aeronautical sectors in our country. It is they who groomed me to take up challenges which are plenty in Design and Development. I am also thankful to all my seniors, peers, and all my colleagues at HAL with whom I worked during this period.

My sincere thanks to other organizations namely Defence Research and Development Organization (DRDO) and its various Labs, Aeronautical Development Agency (ADA), Council of Scientific and Industrial Research (CSIR) and its various Labs; particularly National Aerospace Laboratories (NAL) Bangalore and Central Scientific Instruments Organization (CSIO) Chandigarh, a number of Academic Institutes, other Public and Private sector Industries etc. with whom I had close interaction during my tenure at HAL. My sincere thanks to Indian Military and Civil Certifying Agencies who gave me full support during execution of the projects. I am thankful to our esteemed customers, the Indian Armed and Paramilitary Forces; particularly the Indian Air Force, whose inputs and feedbacks, I always looked forward to in improving our products.

I am extremely grateful to my teams of LCA, IJT, and Jaguar, who made this happen. **A Leader is only as good as the Team he leads**.

My sincere thanks to my current organization, NAL which gave me new career-post superannuation from HAL.

I am also thankful to my colleagues in the ECS group of CSIR-NAL (Messrs. Shankar, Jaidev, Surya and Sapthagiri) who extended help towards preparation of this book.

My particular thanks to Dinakar Aradhya R, Project Associate (CSIR-NAL) who helped me in preparing the manuscript and making a layout of the book.

I am grateful to my family, particularly my wife Kavita, who supported me whole heartedly in this journey and gave valuable inputs.

I am grateful to Abenav who did the design and helped in self-publishing the book.

I sincerely thank Sh. KS Shankara, Corporate Office, HAL for extending support to me in this venture.

My thanks are also due to all my erstwhile senior colleagues, heads of various aerospace organizations including HAL, other professionals I had the privilege to work with, friends, family members, and relatives, who have kindly agreed to join the online book launch event.

In the end, I thank **God Almighty** whose Blessings I always had throughout my life.

'Jai Hind'

Yogesh Kumar

ABOUT THE AUTHOR

Yogesh Kumar, a distinguished alumnus from Punjab Engineering College Chandigarh and IIT Madras, is a renowned technologist in the field of Aircraft and Aircraft Systems.

During his career (all in Research, Design and Development) at Hindustan Aeronautics Ltd. (HAL), a leading Aerospace Organization, he had successfully led a number of major programmes; notable amongst them were;

• India's most prestigious, the Light Combat Aircraft (LCA) programme.

• Development of Intermediate Jet Trainer (IJT) in a record time which has become a benchmark in aircraft development.

• Major Avionics' Upgrade of 'Jaguar' aircraft which was done in record time and which also set a benchmark in concurrent design and production.

• A large number of (over hundred) associated systems and accessories during his tenure as Chief of Design at Lucknow Division of HAL.

All the above aircraft and systems are currently flying.

Author of 23 scientific and technical papers, a book, and over 70 keynote addresses and lectures to reputed organizations, Yogesh Kumar is the proud recipient of following National and International Awards.

- National Aeronautical Prize – 2000 by Aeronautical Society of India (considered the highest in aeronautical field)

- National Acclamation prize - 1988 by Parliamentary Forum on Public Sector Studies

- Raksha Mantri's Award for Excellence - 2003-04 for Import substitution – Given in 2006 by Hon'ble Raksha Mantri (Defence Minister)

- Scientist of the Year Award - 2001 by Gian Chand Jain Memorial Foundation, Ambala Cantt

- Lift time Achievement Award – 2006 by Ambala Scientific Instruments Manufacturers' Association

- Sammana Parta – 2005 by Karnataka Rajyotsava Golden Jubilee Celebration Committee of HAL

- Distinguished Alumnus Award of Excellence – 2018 from Punjab Engineering College (Deemed to be University) Chandigarh

- Padma Shri Nominee (not Awardee) by Ministry of Defence, Govt. of India, 2004

- International Project Management Award – 2005 by International Project Management Association – given to the team led by him. This is considered an **'Oscar'** in Project Management.

- Noel Deerr Gold Medal–1987 by Sugar Technologists' Association of India on paper related to waste-heat recovery-based power plant

Author receiving the prestigious National Aeronautical Prize from Air Chief Marshal A.Y. Tipnis, Chief of Air Staff (January 2001).

Yogesh Kumar is a Fellow, Aeronautical Society of India and was Member, American Institute of Aeronautics and Astronautics and Solar Energy Society of India.

After superannuating from HAL where he was Member of the Board as Director (LCA) in Sept. 2006, he is currently working as Hon' Adviser, Senior Consultant and Specialist Designer in CSIR-NAL; a Govt. of India Aerospace Research Organisation.

He is also an Adjunct Faculty in Punjab Engineering College (Deemed to be University) Chandigarh and Vice Chairman, Design Division of Aeronautical Society of India.

He is currently advising CSIR-NAL on its civil aircraft programmes as below;

- The Light Transport Aircraft 'Saras'- development and technology transfer to Industry Partner.

- The Indian Regional Transport Aircraft- preparation of Project Reports, Design of Systems etc.

Contact Information:

E-mail: yogesh_hal@yahoo.co.in

Mobile No.: +91 9845180134,

www.linkedin.com/in/yogeshkumarhal

REFERENCES

During preparation of the book, following references used to illustrate the points by me are gratefully acknowledged;

1. Previous book of mine titled;

    "Lead and Execute –

    The Art of Managing Large Scale Projects"

    It was launched in Dec. 2014 and within two years, over 600 copies were picked up by various organizations.

    It is now available on Amazon platform at the following link;

    https://www.amazon.com/dp/B089GP89Y4

2. Quotable Quotes, Statements made by various scholars and eminent personalities.

3. Personal notes prepared during various management courses attended by me in reputed organizations during my career at HAL.

Author and his family with former President Dr. Abdul Kalam at Rashtrapati Bhavan

(14 Jan.2003)

REMINISCENES

Photographs* of VVIP visits
(*Courtesy HAL)

Photographs of Book Launch
Dec. 2014

PHOTOGRAPHS OF VVIP VISITS

	*On occasion of the first flight of the **Light Combat Aircraft** (TD-1) on 4 January 2001*
*During the visit of Shri George Fernandes on occasion of the inaugural flight of **IJT** on 21 March 2003*	
	*Shri Yogesh Kumar receiving a memento from Hon'ble Raksha Mantri Shri George Fernandes on occasion of the inaugural flight of the Intermediate Jet Trainer **(IJT)***
*With Hon'ble Prime Minister, Shri AB Vajpayee on occasion of the naming ceremony of **Light Combat Aircraft** on 4 May 2003*	

PHOTOGRAPHS OF VVIP VISITS

Visit of Hon'ble Raksha Mantri, Shri Pranab Mukherjee to HAL on 13 June 2004

On occasion of the visit of Hon'ble Raksha Mantri, Shri Pranab Mukherjee to HAL on 13 June 2004.

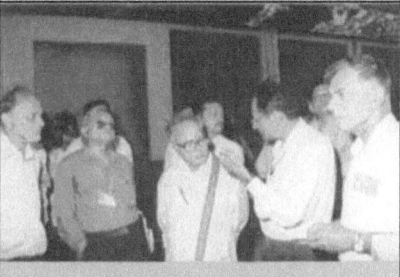

*Dr. Manmohan Singh, Hon'ble Prime Minister of India being presented with a model of the **Tejas** Light Combat Aircraft during his visit to HAL on 12 February 2005*

Dr. Manmohan Singh being presented a model of the Intermediate Jet Trainer during his visit to HAL on 12 February 2005.

BOOK LAUNCH FUNCTION- 18 DECEMBER. 2014

www.ingramcontent.com/pod-product-compliance
Lightning Source LLC
Chambersburg PA
CBHW020438220526
45464CB00002B/757